*Automation of Cytogenetics*

C. Lundsteen    J. Piper (Eds.)

# Automation of Cytogenetics

With 103 Figures

Springer-Verlag Berlin Heidelberg New York
London Paris Tokyo Hong Kong

Claes Lundsteen, M.D., Ph. D.
Rigshospitalet, Section of Clinical Genetics 4032
Department of Obstetrics and Gynaecology
DK-2100 Copenhagen Ø, Denmark

Dr. Jim Piper
MRC Human Genetics Unit, Western General Hospital, Crewe Road
GB-Edinburgh EH4 2XU, Scotland

Publication No. EUR 11968 of the Scientific and Technical Communication Unit, Commission of the European Communities, Directorate-General Telecommunications, Information Industries and Innovation, Luxembourg

ISBN 3-540-51105-9 Springer-Verlag Berlin Heidelberg New York
ISBN 0-387-51105-9 Springer-Verlag New York Berlin Heidelberg

Library of Congress Cataloging-in-Publication Data. Automation of cytogenetics. Consists of papers from a series of Workshops on the Automation of Cytogenetics, held in Berlin 1986, Cortona 1987, and Langollen 1988, sponsored by the Commission of the European Communities. Bibliography: p. Includes index. 1. Human chromosomes–Analysis–Automation–Congresses. 2. Human cytogenetics–Automation–Congresses. 3. Human chromosome abnormalities–Diagnosis–Automation–Congresses. I. Lundsteen, Claes. II. Piper, J. (Jim), 1948-. III. Workshop on the Automation of Cytogenetics. IV. Commission of the European Communities. [DNLM: 1. Automation–congresses. 2. Chromosomes–analysis–congresses. 3. Cytogenetics–congresses. QH 605 A939] RB155.6.A97 1989 611'.01816 89-11298 (U.S. : alk. paper)

© ECSC-EEC-EAEC, Brussels-Luxembourg, 1989
Printed in Germany

Legal Notice
Neither the Commission of the European Communities nor any person acting on behalf of the Commission is responsible for the use which might be made of the following information.

Printing: Druckhaus Beltz, 6944 Hemsbach/Bergstr.
Binding: J. Schäffer GmbH & Co. KG., D-6718 Grünstadt
2156/3150-543210 – Printed on acid-free paper

# Preface

This book complements the successful series of Workshops on the Automation of Cytogenetics which have been sponsored for more than a decade by the European Community (EC). The contributed papers all arise from presentations at one of the last three such Workshops, in Berlin (1986), Cortona (1987) and Langollen (1988). Collaborative cytogenetics automation research and development activities in Europe are now supported by the "Concerted Action in Automation of Cytogenetics" (CAACG), as one of the activities of the EC's COMAC-BME committee which supervises the coordination of research in biomedical engineering within the Medical Technology Development target (project no. II.1.1/13). The structure of EC programme for medical and health research is described on the following pages.

The book presents the state of the art in the automation of cytogenetics, comprising 20 papers by researchers, developers and users of cytogenetics automation. Topics covered include evaluations of complete image-based chromosome analysis systems which have resulted from the successful collaboration between European researchers and industry, current active research into new image analysis techniques (with particular emphasis on techniques for automated segmentation of chromosome clusters), the development of systems for chromosome aberration scoring, automatic specimen preparation, and the use of flow cytometry for chromosome measurement and analysis.

An extensive bibliography covering publications in all aspects of chromosome analysis automation is included; we expect that maintaining and circulating this bibliography will be one extremely useful activity of CAACG, and so we ask all our readers to send us new titles in this field, and to point out any earlier papers that we have overlooked.

We should like to thank all the contributors for submitting their papers, Thorkil Boisen for his work on the subject index, Dr. H.-U. Daniel of Springer-Verlag for his continued interest and assistance, and the Commission of the European Communities for sponsoring both the Workshops and this book.

Claes Lundsteen
Jim Piper

# Medical and Health Research Programme of the EC

## Biomedical Engineering in the European Community

The involvement of the European Community (EC) in the field of Medical and Health Research started in 1978 with the first Programme which contained three projects. Since then, it has steadily expanded and it will include around 120 projects by the end of the fourth Programme (1987-1991).

The *general goal* of the programme is clearly to contribute to a better quality of life by improving health, and its *distinctive feature* is to strengthen European collaboration in order to achieve this goal.

The main objectives of this collaboration are:

- increase the *scientific efficiency* of the relevant research and development efforts in the Member States through their gradual coordination at Community level following the mobilization of the available research potential of national programmes, and also their *economic efficiency* through sharing of tasks and strengthening the joint use of available health research resources,

- *improve scientific and technical knowledge* in the research and development areas selected for their importance to all Member States, and promote its *efficient transfer into practical applications*, taking particular account of potential industrial and economic developments in the areas concerned,

- *optimize the capacity and economic efficiency of health care efforts* throughout the countries and regions of the Community.

The current programme consists of six research targets. Four are related to major health problems: **Cancer, AIDS, Age-Related Problems,** and **Personal Environment and Life-Style Related Problems**; two are related to health resources: **Medical Technology Development** and **Health Services Research**.

Funds are provided by the Community for relevant "concerted action" activities which consist of research *collaboration and coordination* in EC Member States and/or in other European participant countries. *Networks* of research institutes can be set up and supported by means of meetings, workshops, short-term staff exchanges/visits to other countries, information dissemination and so on; centralized facilities such as data banks, computing, and preparation and distribution of reference materials can also be funded. The funds are *not* direct research grants; the institutes concerned must fund the research activities carried out within their own countries – it is the international coordination activities which are eligible for Community support. Each such research network is placed under the responsibility of a *Project Leader* chosen from among the leading scientists in the network, with the assistance of a *Project Management Group* representing the teams participating in the network.

The Commission of the European Communities is assisted in the execution of this Programme by a Management and Coordination Advisory Committee (CGC – Medical and Health Research), and by Concerted Action Committees (COMACs) and Working Parties, composed of representatives and of scientific experts respectively, designated by the competent authorities of the Member States.

*Other European Countries*, not belonging to the EC but participating in COST (Cooperation on Science and Technology) may take part in the Programme.

The present work was conducted according to the advice of COMAC-BME which supervises the coordination of research in biomedical engineering (BME) within the Medical Technology Development target.

More information may be obtained from:

Commission of the European Communities

Directorate General XII-F-6

200 Rue de la Loi

B–1049 Brussels

Tel.: +32-2-235.00.34 / Telex: COMEU B 21877

Telefax: +32-2-235.01.45; +32-2-236.20.07

# Contributors

L. Amory — ESAT-MI2, Department of Electrical Engineering, University of Leuven, de Croylaan 52, 3030 Heverlee, Belgium

Jacob A. Aten — Laboratory of Radiobiology, University of Amsterdam, AMC Meibergdreef 9, 1105 AZ Amsterdam, The Netherlands

Richard Baldock — MRC Human Genetics Unit, Western General Hospital, Crewe Road, Edinburgh EH4 2XU, Scotland, UK

P. Battaglia — Cattedra di Istologia ed Embriologia Generale, Universita degli Studi di Modena, Via del Pozzo 71, Modena, Italy

J. Bille — Institute of Applied Physics I, University of Heidelberg, Albert-Überle-Str. 3-5, 6900 Heidelberg, Federal Republic of Germany

R. Bolzani — Cattedra di Istologia ed Embriologia Generale, Universita degli Studi di Modena, Via del Pozzo 71, Modena, Italy

P. Bonton — Laboratoire d'Electronique, Université Clermont II, B.P. 45, 63170 Aubiere, France

Michael C. Cimino — Section of Human Genetics, Department of Obstetrics and Gynecology, Northwestern University Medical School, Chicago, IL, USA

Maimon M. Cohen — Division of Human Genetics, Department of Obstetrics and Gynecology, Department of Pediatrics, University of Maryland Medical Center; Medical Biotechnology Center of the Maryland Biotechnology Institute; Baltimore, MD, USA.

A. Cooke — Duncan Guthrie Institute of Medical Genetics, Royal Hospital for Sick Children, Yorkhill, Glasgow G3 8SJ, Scotland, UK

Cristoph Cremer — Institute of Applied Physics I, University of Heidelberg, Albert-Überle-Str. 3-5, 6900 Heidelberg, Federal Republic of Germany

Thomas Cremer — Institute of Anthropology and Human Genetics, Im Neuenheimer Feld 328, 6900 Heidelberg, Federal Republic of Germany

Michael G. Daker — Paediatric Research Unit, Division of Medical and
Molecular Genetics, United Medical and Dental Schools of
Guy's and St. Thomas's Hospitals, 8th floor, Guy's Tower,
London SE1 9RT, UK

Eduardo Diaz — Institute of Applied Physics I, University of Heidelberg,
Albert-Überle-Str. 3-5, 6900 Heidelberg, Federal Republic of
Germany

Alan Dyer — Department of Community Health and Preventive Medicine,
Northwestern University Medical School, Chicago, IL, USA

Judith A. Fantes — MRC Human Genetics Unit, Western General Hospital,
Crewe Road, Edinburgh EH4 2XU, Scotland, UK

Steve Farrow — MRC Human Genetics Unit, Western General Hospital,
Crewe Road, Edinburgh EH4 2XU, Scotland, UK

A. Forabosco — Cattedra di Istologia ed Embriologia Generale, Universita
degli Studi di Modena, Via del Pozzo 71, Modena, Italy

J.M. García-Sagredo — Medical Genetics Department, Hospital Ramón y
Cajal, 28034 Madrid, Spain

C.R.G. van der Geest — Mechanics Department, Sylvius Laboratories,
University of Leiden, Wassenaarseweg 62, 2333 AL Leiden,
The Netherlands

A. Geneix — Laboratoire de Cytogénétique, Faculté de Médecine, B.P. 38,
63001 Clermont Ferrand Cédex, France

Tommy Gerdes — Section of Clinical Genetics, Department of Obstetrics and
Gynecology; Department of Medical Engineering; Rigshospitalet,
Copenhagen, Denmark

G.W. Gerrese — Mechanics Department, Sylvius Laboratories, University of
Leiden, Wassenaarseweg 62, 2333 AL Leiden, The Netherlands

James Graham — Wolfson Image Analysis Unit, Department of Medical
Biophysics, University of Manchester, Oxford Road, Manchester,
UK

E. Granum — Institute of Electronic Systems, Aalborg University,
9000 Aalborg, Denmark

J. Gregor — Institute of Electronic Systems, Aalborg University,
9000 Aalborg, Denmark

Daryll K. Green — MRC Human Genetics Unit, Western General Hospital,
Crewe Road, Edinburgh EH4 2XU, Scotland, UK

Frans C.A. Groen — Faculty of Mathematics and Computer Science,
University of Amsterdam, Amsterdam, The Netherlands

L. Grouche — Laboratoire d'Electronique, Université Clermont II, B.P. 45, 63170 Aubiere, France

Michael Hausmann — Institute of Applied Physics I, University of Heidelberg, Albert-Überle-Str. 3-5, 6900 Heidelberg, Federal Republic of Germany

Jutta Hetzel — Institute of Anthropology and Human Genetics, Im Neuenheimer Feld 328, 6900 Heidelberg, Federal Republic of Germany

P. Huygue — Laboratoire d'Electronique, Université Clermont II, B.P. 45, 63170 Aubiere, France

Edmund C. Jenkins — Department of Cytogenetics, Institute for Basic Research, New York State Office of Mental Retardation and Developmental Disabilities, Staten Island, NY, USA

Liang Ji — MRC Human Genetics Unit, Western General Hospital, Crewe Road, Edinburgh EH4 2XU, Scotland, UK

G. Korthof — Department of Human Genetics, Sylvius Laboratories, University of Leiden, Wassenaarseweg 62, 2333 AL Leiden, The Netherlands

Ton K. ten Kate — Faculty of Medicine, Free University, Amsterdam, The Netherlands

Dean Kowal — Section of Human Genetics, Department of Obstetrics and Gynecology, Northwestern University Medical School, Chicago, IL, USA

Th. Lörch — Institute for Radiation Hygiene, Federal Health Office, Neuherberg, Federal Republic of Germany

D.C. Lloyd — National Radiological Protection Board, Chilton, Didcot, Oxon, UK.

Claes Lundsteen — Section of Clinical Genetics, Department of Obstetrics and Gynecology, Rigshospitalet, Copenhagen, Denmark

A. Oosterlinck — ESAT-MI2, Department of Electrical Engineering, University of Leuven, de Croylaan 52, 3030 Heverlee, Belgium

Jan Maahr — Section of Clinical Genetics, Department of Obstetrics and Gynecology, Rigshospitalet, Copenhagen, Denmark

P. Malet — Laboratoire de Cytogénétique, Faculté de Médecine, B.P. 38, 63001 Clermont Ferrand Cédex, France

Sheri Maremont — Section of Human Genetics, Department of Obstetrics and Gynecology, Northwestern University Medical School, Chicago, IL, USA

Alice O. Martin — Section of Human Genetics, Department of Obstetrics and Gynecology, Northwestern University Medical School, Chicago, IL, USA

Brian H. Mayall — Laboratory for Cell Analysis, University of California San Francisco, San Francisco, CA, USA

Mark McKie — MRC Human Genetics Unit, Western General Hospital, Crewe Road, Edinburgh EH4 2XU, Scotland, UK

R.D. McKinney — Section of Human Genetics, Department of Obstetrics and Gynecology, Northwestern University Medical School, Chicago, IL, USA

P.L. Pearson — Department of Human Genetics, Sylvius Laboratories, University of Leiden, Wassenaarseweg 62, 2333 AL Leiden, The Netherlands

Jim Piper — MRC Human Genetics Unit, Western General Hospital, Crewe Road, Edinburgh EH4 2XU, Scotland, UK

Amy Rissman — Section of Human Genetics, Department of Obstetrics and Gynecology, Northwestern University Medical School, Chicago, IL, USA

Roelof Roos — Pattern Recognition Group, Faculty of Applied Physics, Delft University of Technology, Delft, The Netherlands.

Denis Rutovitz — MRC Human Genetics Unit, Western General Hospital, Crewe Road, Edinburgh EH4 2XU, Scotland, UK

Howard Sabrin — Section of Human Genetics, Department of Obstetrics and Gynecology, Northwestern University Medical School, Chicago, IL, USA

Michael Shaunnessey — Section of Human Genetics, Department of Obstetrics and Gynecology, Northwestern University Medical School, Chicago, IL, USA

Joe Leigh Simpson — Section of Human Genetics, Department of Obstetrics and Gynecology, Northwestern University Medical School, Chicago, IL, USA

J. Snellings — ESAT-MI2, Department of Electrical Engineering, University of Leuven, de Croylaan 52, 3030 Heverlee, Belgium

Margaret Stark — MRC Human Genetics Unit, Western General Hospital, Crewe Road, Edinburgh EH4 2XU, Scotland, UK

G. Stephan — Institute for Radiation Hygiene, Federal Health Office, Neuherberg, Federal Republic of Germany

P. Suetens — ESAT-MI2, Department of Electrical Engineering, University of Leuven, de Croylaan 52, 3030 Heverlee, Belgium

M.G. Thomason — Computer Science Department, University of Tennessee, Knoxville, TN 37996, USA

Simon Towers — MRC Human Genetics Unit, Western General Hospital, Crewe Road, Edinburgh EH4 2XU, Scotland, UK

G. Vletter — Mechanics Department, Sylvius Laboratories, University of Leiden, Wassenaarseweg 62, 2333 AL Leiden, The Netherlands

Lucas J. van Vliet — Pattern Recognition Group, Faculty of Applied Physics, Delft University of Technology, Delft, The Netherlands.

Albert M. Vossepoel — Department of Medical Informatics, University of Leiden, Wassenaarseweg 62, 2333 AL Leiden, The Netherlands

J. Vrolijk — Department of Human Genetics, Sylvius Laboratories, University of Leiden, Wassenaarseweg 62, 2333 AL Leiden, The Netherlands

C. Wittler — Institute for Radiation Hygiene, Federal Health Office, Neuherberg, Federal Republic of Germany

Q. Wu — ESAT-MI2, Department of Electrical Engineering, University of Leuven, de Croylaan 52, 3030 Heverlee, Belgium

Ian T. Young — Pattern Recognition Group, Faculty of Applied Physics, Delft University of Technology, Delft, The Netherlands.

# Contents

Chromosome Aberration Detection with Hybridized DNA Probes:
Digital Image Analysis and Slit Scan Flow Cytometry (C. Cremer,
M. Hausmann, E. Diaz, J. Hetzel, J.A. Aten, T. Cremer) .............. 123

Part IV      Automatic Preparation of Cytogenetic Specimens

An Automated System for the Culturing and Harvesting of Human
Chromosome Specimens (J. Vrolijk, G. Korthof, G. Vletter,
C.R.G. van der Geest, G.W. Gerrese, P.L. Pearson).................... 135

Evaluation and Development of a System for Automated Preparation
of Blood Specimens for Cytogenetic Analysis (A.O. Martin,
M. Shaunnessey, H. Sabrin, S. Maremont, A. Dyer, M.C. Cimino,
A. Rissman, R.D. McKinney, M.M. Cohen, E.C. Jenkins,
D. Kowal, J.L. Simpson) ............................................. 149


Part V       Automatic Chromosome Segmentation

Decomposition of Overlapping Chromosomes (L. Ji) ................... 177

Resolution of Composites in Interactive Karyotyping
(J. Graham) ......................................................... 191

Separation of Touching Chromosomes (A.M. Vossepoel)................ 205

Model-Based Contour Analysis in a Chromosome Segmentation System
(Q. Wu, J. Snellings, L. Amory, P. Suetens, A. Oosterlinck)............ 217


Part VI      Chromosome Classification

On the use of Automatically Inferred Markov Networks for
Chromosome Analysis (E. Granum, M.G. Thomason, J. Gregor) ........ 233

Density Profiles in Human Chromosome Analysis
(A. Forabosco, P. Battaglia, R. Bolzani) ............................. 253

Cytogenetic Analysis by Automatic Multiple Cell Karyotyping
(C. Lundsteen, T. Gerdes, J. Maahr) ................................. 263

Towards a Knowledge-Based Chromosome Analysis System
(J. Piper, R. Baldock, S. Towers, D. Rutovitz)........................ 275

Bibliography - Cytogenetics Automation 1975-88 ..................... 295

Subject Index........................................................ 307

# Introduction

Denis Rutovitz

This book marks a quarter century since the automation of chromosome analysis was first proposed, and a decade of cooperation in its pursuit within the European Community. Since there are now numerous machines in use in cytogenetics laboratories in many different countries, it is apparent that the goal proposed has, at least in part, been achieved. Since most of the machines in use world-wide are manufactured by European companies participating in the cooperative action, it is apparent too that the Community's interest has been in some measure repaid.

From conception to fruition was a long gestation. In the evolution of the subject one can mark three phases. In the late 60's and early 70's there was a heady rush of discovery in image analysis in general, and in its chromosome arm in particular: minor discovery, by general scientific standards, but satisfying enough to the participants none the less. Examples of such discoveries or inventions which were closely associated with chromosome work: syntax-directed analysis of object boundaries (Ledley, 1964), the Hilditch skeleton (Hilditch, 1969), the Prewitt operators (Prewitt, 1970), object oriented image processing (Rutovitz, 1968), the Fidac film digitiser (Ledley et al, 1965), optical processing of microscope images in the Fourier domain (Wald and Preston, 1968), densitometric analysis and local resegmentation of images (Mendelsohn et al, 1966). This period saw the first systems for digitisation, classification, interaction and production of hard copy karyotypes (albeit on line printers or in outline on plotters), and the first metaphase finders.

But, despite the excitement, the systems delighted their inventors more than their potential users, the cytogeneticists. Looking back, it now seems obvious that the performance of the hardware of the time could never justify its price in a cytogenetics laboratory; and that the software, while undoubtedly interesting, had a very long way to go to achieve a worthwhile level of reliability and usability.

There followed a long period of consolidation, in which few new algorithms or inventions appeared, but in which there were serious attempts at the design and prototyping of complete systems, with computer controlled microscope stages and digitisers, metaphase finding capability, adequate computing power (sometimes obtained by special hardware or the use of parallelism), and screen-based interaction, at Edinburgh (Green and Neurath, 1974, Green at al, 1977), Jet Propulsion Laboratory (Castleman et al, 1975), Tufts New England Medical

**Automation of Cytogenetics**  Editors: C. Lundsteen J. Piper
© Springer-Verlag Berlin Heidelberg New York 1989

Centre (Neurath et al, 1975) and elsewhere. In this period also the emphasis shifted finally from film toward microcope-based interactive systems.

The third, and present phase began with the initiative of Lundsteen and Philip of the Rigshospitalet in Copenhagen, in commissioning Joyce-Loebl to build what was originally conceived of as a commercial version of the JPL machine. Joyce-Loebl were already in the image analysis field with their "Magiscan" device designed by C. Taylor of the Wolfson Image Analysis Unit of Manchester University (Graham and Taylor, 1980). For the first time this brought a chromosome analysis system into a working cytogenetics laboratory: the child conceived 20 years before was delivered at last. One understands that the labour pains were severe, and of several years duration; but when the world observed that the parents were, finally, proud and pleased, it became firmly established that a machine-aided chromosome analysis system was a desirable and proper purchase for a cytogenetics laboratory (Lundsteen et al, 1983).

The Copenhagen developments resulted in a wave of new interest, especially in Europe. One of the results was continuation of European Community funding for a series of workshops on the subject. The first European workshops were held in Leuven in 1976 and Leiden in 1977 under the auspices of the Boerhave foundation, but from then onwards, beginning with a meeting in Copenhagen in 1978, they were underwritten by the Community. This steady support culminated in the establishment a "Concerted Action" in 1988. The papers in this volume have all been informally presented at these workshops.

On the industrial side the effect was dramatic: many companies followed Joyce-Loebl's lead and began production of systems, for the most part in association with a group with a long history of research on the subject. Thus the Leitz "Leytas" system was inspired by the Mathemetical Morphology theories of Serra (1982) and the interest of Peter Pearson and the Leiden Image Analysis group (Vrolijk et al, 1980). Image Recognition Systems in Warrington U.K. implemented the ideas of the Edinburgh M.R.C. group (Lloyd et al, 1987); Perceptive Systems Inc., founded by Ken Castleman, based their design on his successful JPL work; Ledley and Lubs launched "Metachrome", a distant successor to the Fidac. Others, such as Kontron in Munich, implemented new systems on the familiar lines.

Consequently there is now a body of cytogeneticists who have accumulated several years experience of using machines, and who can report on the benefits and the drawbacks of the systems on a solid and well established basis; and most believe that the benefits outway the drawbacks.

Does that mean that the research task is done, and that the manufacturers and the laboratories are now best left to satisfy each others' further requirements, while the engineers and scientists of the image analysis community look to new targets elsewhere? Unfortunately, no. If we ask what has really been provided so far, we see that it amounts to only a modicum of assistance to the human operator. The amount of time that is required to produce karyotypes has been somewhat reduced, as has been the amount of unskilled work that needs to be done. It no longer necessary to have skilled technicians cutting up photographs with a pair of scissors and arranging chromosomes with adhesive paste, and that is certainly a good thing. In other words, apart from metaphase finding, essentially we have provided a helpful computer graphics system, with very limited interpretive ability: while the automatic chromosome analysis infant is alive and well, it has a long way to go to reach maturity. Nevertheless we can be confident that the industrial and user communities will now ensure that the more mundane, but essential, underpinning requirements will be looked after very effectively. These include questions of slide handling, of image acquisition, of operator interaction and generally of system provision. For the researcher, it is therefore an opportune time to try to push forward our understanding of these images and their processing problems. Because of the continuing remarkable increase in power and decrease in cost of computer processing units and memories, it is at last realistic to plan a laboratory instrument which uses complex procedures needing the support of sophisticated machines, languages and operating systems. In particular it is an appropriate time to try to apply, systematically, the methods of control of complex processes by access to knowledge bases which have become established in artificial intelligence, and are ripe for exploitation in our field.

There are many challenges still before us, but since there is now a good base of installed semi-automated equipment, we can be sure that research achievements will be translated rapidly into real working systems in the user community. We will not have to wait 20 years for the next generation of methods and strategies to be taken up!

What are the advances which we can really look forward to? I would put forward four as being the most significant. The first of these is an almost fully automated aberration scoring system such as is being developed by the Heidelberg and Edinburgh groups (Bille et al, 1987, Lörch, this volume, Piper et al, 1988). If successful these will be highly significant developments. It has been shown by several independent investigators (Lundsteen et al, 1987, Martin and Unger, 1985, Daker, this volume) that present karyotyping/metaphase finding machines increase operator throughput by about 50%, at most. In contrast aberration scoring systems are expected to increase throughput, compared to that of the unassisted operator, by a factor of over 10:1 in elapsed time, and over 100:1 in terms of operator's time: this would be a very real automation. Closely associated and/or parallel developments are on the one hand automatic digitisation and selection of cells for analysis (Stark et al, this volume), and on the other developments in segmentation and classification procedures which will greatly reduce the amount of operator interaction required in a karyogram (Graham, this volume, Ji, 1989, Ji, this volume, Vossepoel, this volume, Wu, this volume). Other more specific developments which we can now seriously anticipate are methods of formation of composite karyotypes which fully automatically retrieve restricted ranges of abnormalities such as aneuploid cells

(Carothers et al, 1983). Systems of that type might not be suitable for general clinical purposes but may well have application in research systems, for example in drug evaluation using animal tests. Lastly, developments in retrieval of band-patterns, and their analysis, will hopefully soon lead to the automatic extraction of meaningful band-pattern descriptions, replacing the inexact and global methods to which we are constrained today (Wienberg, 1989, Granum et al, this volume).

There is a further way to go, and the European Concerted Action may have an important contribution to make towards the achievement of rapid progress. Meanwhile, we hope that readers of this book will find it a helpful guide to what has already passed, and a useful indication of what may soon come to pass.

# References

1.  Bille J, Loerch T, Stephan G, Wittler C: Automated cytogenetics dosimetry. Proc IEEE 9th Annual Conf. Eng. in Med and Biol. Boston Nov. 13-16 (1987)
2.  Carothers AD, Rutovitz D, Granum E: An Efficient Multiple-Cell Approach to automatic aneuploidy screening. Anal Quant Cytol 5:194-200 (1983)
3.  Daker MG. A Cost Evaluation of Magiscan 2 (this volume)
4.  Graham J, Taylor CJ. Automated chromosome analysis using the Magiscan image analyser. Anal Quant Cytol 2:237-242 (1980)
5.  Graham J. Resolution of Composites in Interactive Karyotyping (this volume)
6.  Granum E, Thomason MG, Gregor J. On the use of Automatically Inferred Markov Networks for Chromosome Analysis (this volume)
7.  Green DK, Neurath PW. The design, operation and evaluation of a high speed automatic metaphase finder. J Histochem Cytochem 22:531-535 (1974)
8.  Green DK, Bayley R, Rutovitz D. A cytogeneticists's microscope. Microscopia Acta 79:237-245 (1977)
9.  J Hilditch. Linear skeletons from square cupboards. In: Meltzer B, Michie D (eds) Machine Intelligence 4 (Edinburgh University Press, 1969) pp.403-420
10. Ji L. Decomposition of Overlapping Chromosomes (this volume)
11. Ji L. Intelligent splitting in the chromosome domain. Patt Recog (in press, 1989)
12. Ledley RS. High speed automatic analysis of biomedical pictures. Science 146:216-223 (1964)
13. Ledley RS, Rotolo LS, Golab TJ, Jacobsen JD, Ginsberg MD, Wilson TB. FIDAC: film input to digital automatic computer and associated syntax-directed pattern recognition programming system. Optical and Electro-optical Information Processing (M.I.T. Press, Cambridge, MA, 1965) pp.591-614
14. Lloyd D, Piper J, Rutovitz D, Shippey G: . Multiprocessing interval processor for automated cytogenetics. Appl Optics 26:3356-3366 (1987)
15. Lörch T, Wittler C, Stephan G, Bille J. An Automated Chromosome Aberration Scoring System (this volume)
16. Lundsteen C, Gerdes T, Philip J, Graham J, Pycock D: An interactive system for chromosome analysis. Tests of clinical performance. Proc 3rd Scan Conf Image Analysis, Copenhagen (1983) pp.392-397.
17. Lundsteen C, Gerdes T, Maahr J, Philip J: Clinical performance of a routine system for semi-automated chromosome analysis. Am J Hum Genet 41:493-502 (1987)
18. Martin AO, Unger N. Automated cytogenetics, Comtemporary Ob-Gyn Technology (1985)
19. Mendelsohn ML, Conway TJ, Hungerford DA, Kolman WA, Perry BH, Prewitt JMS. Computer-oriented analysis of human chromosomes, I. Photometric estimation of DNA content. Cytogenetics 5:223-242 (1966)

20. Neurath PW, Gallus G, Horton JB, Selles W: Automatic karyotyping: Progress, Perspectives and Economics. In: Mendelsohn ML (ed) Automation of Cytogenetics, Asilomar Workshop. (Lawrence Livermore Laboratory Technical report CONF-751158, 1975) pp.17-26.
21. Piper J, Towers S, Gordon J, Ireland J, McDougall D. Hypothesis combination and context sensitive classification for chromosome aberration scoring. In: Gelsema ES, Kanal LN (eds) Pattern Recognition and Artificial Intelligence (Elsevier, Amsterdam, 1988) pp.449-460
22. Prewitt, JMS. Object enhancement and extraction In: Lipkin BS, Rosenfeld A (eds) Picture Processing and Psychopictorics (Academic Press, New York, 1970) pp.75-149
23. Rutovitz D. Data structures for operations on digital images. In: Cheng GC, Ledley RS, Pollock DK, Rosenfeld A (eds) Pictorial Pattern Recognition (Thompson, Washington, DC, 1968) pp.105-133
24. Serra J. Image Analysis and Mathematical Morphology (Academic Press, New York, 1982)
25. Stark M, Farrow S, McKie M, Rutovitz D. Automatic High Resolution Digitisation of Metaphase Cells for Aberration Scoring and Karyotyping (this volume)
26. Vossepoel AM. Separation of Touching Chromosomes (this volume)
27. Vrolijk J, Pearson PL, Ploem JS: Leytas, a system for the processing of microscopic images. Anal Quant Cytol 2:41-48 (1980)
28. Wald N, Preston K. Automatic screening of metaphase spreads for chromosome analysis. In: Ramsay DM (ed) Proceedings of the Conference on Image Processing in Biological Science. UCLA Forum in Medical Sciences no. 9 (University of California Press, Berkeley, CA, 1968) pp.9-34
29. Wienberg J. Poster presentation at 10th European Workshop on Automated Cytogenetics (Llangollen, 1988).
30. Wu Q, Snellings J, Amory L, Suetens P, Oosterlinck A. A Polygonal Approximation Approach to Model-Based Contour Analysis in a Chromosome Segmentation System (this volume)

# Part I

# Automatic Chromosome
# Aberration Scoring

# Automated aberration scoring : the requirements of an end-user

D.C. Lloyd

**Summary**

This paper briefly reviews requirements of a laboratory intending to invest in a system for automatic analysis of non-constitutional aberrations for mutagen studies. The essential endpoints to be measured are classical aberrations, sister chromatid exchanges and micronuclei. Particular emphasis is given to discussing scoring for unstable translocations such as the dicentric. The opportunity is also taken to emphasise the growing interest in the micronucleus assay.

## 1. Introduction

In 1984 the National Radiological Protection Board purchased the front end of a Cytoscan instrument. This is a metaphase finder but without the capability to karyotype. After a period of familiarisation and then fairly rigorous evaluation of the performance of the instrument (Finnon et al 1986), it is now in constant use as an aid to the routine workload of the laboratory examining metaphases of human peripheral blood lymphocytes for chromosomal damage induced by ionising radiation. I thus comment as an "end-user" but so far only of a metaphase finder with the hope that eventually the instrument will be replaced or improved by the addition of facilities for automated aberration scoring.

Metaphase finding has been achieved in a number of systems but to date automated scoring has not been successfully developed. However, a few groups, notably in Heidelberg and Edinburgh, are at various stages along the way. In the case of development of the Cytoscan, most effort during the mid-1980s was concentrated on perfecting semi-automated karyotyping. This was achieved so that some effort has become available to be deployed towards developing the system for aberration analysis.

## 2. Classical aberrations

Karyotyping will of course detect aberrant chromosomes, but it is really designed to pick up the inheritable constitutional abnormalities that accompany clinically significant syndromes or diseases. Research on radiation induced

**Automation of Cytogenetics**   Editors: C. Lundsteen J. Piper
© Springer-Verlag Berlin Heidelberg New York 1989

chromosomal damage involves a major difference in that usually one is looking for non-constitutional changes, such as dicentrics, that differ from cell to cell and which in many instances are present only at a very low frequency; they often conform to the Poisson distribution of "rare random events". This inevitably leads to the need to examine many cells, hundreds or even thousands per sample, in order to produce a quantitative result carrying acceptably small statistical uncertainty. It was this that made it attractive to purchase an instrument that finds and ranks the quality of the metaphases and then presents them centralised in focus at high magnification for the technician to analyse by eye for the presence of aberrations.

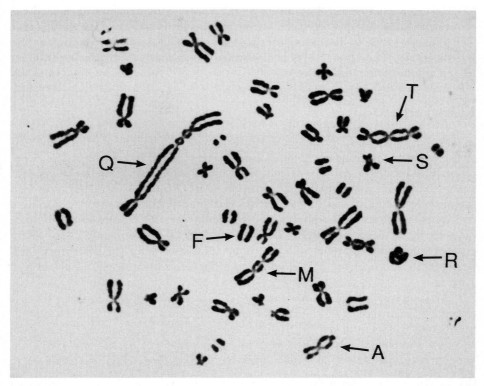

**Fig. 1.** A first division metaphase (M1) uniformly stained and containing a large number of aberrations: M - metacentric dicentric, A - singly acrocentric dicentric, S - short dicentric, R - ring, T - tricentric, Q - quadracentric, F - fragment

When scoring with a conventional microscope a technician can usually maintain a rate of about 250 metaphases per day, unless the material is highly damaged. A faster rate leads to inaccuracy due to fatigue. After the initial evaluation of the Cytoscan was completed in our laboratory, data have been accumulated over a period of a few years while the instrument has been in routine use. It has been found that the output of the technician has approximately trebled by comparison with scoring on a conventional microscope. One would hope that automation would considerably speed up the aberration scoring rate, even if, as with karyotyping, it still requires operator assistance. A slower automatic system that is entirely "hands off" may also be worthwhile because it could be set to work continuously day and night without fatigue. Realistically, what is likely to emerge is a system that can work unaided for long periods but which at the end of the list presents a proportion of objects for confirmation by a technician. Indeed, current cytogenetic practice is such that one would not consider any system that classifies a chromosome as being abnormal without it having been checked by an operator. For an automated scoring system, operator intervention for every cell to deal with false positives would not be economically feasible. This is because in contrast to karyotyping it is unlikely that an interactive aberration analysis could be significantly faster than conventional scoring. However, if such an instrument could successfully deal unaided with the considerable majority of cells containing no aberrations with the assurance that the false negative rate is low, then this would be a major advance.

When presented by the metaphase finder with the cell for visual scoring for unstable aberrations such as dicentrics, rings and fragments, a skilled technician takes approximately 30 seconds to complete the analysis of an undamaged cell. If there are aberrations present, the time required may be longer. With conventional scoring the first step is to count the number of objects in the cluster and if there is less than 46 the metaphase is usually rejected. Step two is to examine the cluster, looking for aberrant chromosomes and to correlate them, ie. a dicentric must be accompanied by a fragment but the cluster count remains at 46; a ring and its fragment increases the count to 47; any fragments not associated with an unstable translocation should correlate with a cluster count in excess of 46.

When considering automation, the necessity for cluster counting should be reviewed. It may be just as accurate to use the dicentric count from all metaphases irrespective of their completeness because the likelihood of an object being lost as an artefact of fixation and slide preparation is presumably unrelated to whether it is a, mono or dicentric. One might compensate for the missing chromosomes from the incomplete spreads, simply by scoring more metaphases and expressing the result as "dicentrics per chromosome" rather than the more conventional "per cell". This is something that will need to be determined by experiment when an automated system becomes available. The importance of cluster counting as a double check of the identification of aberrations is likewise amenable to confirmation by experiment. If one can dispense with counting that inevitably has to deal with a few touching and overlapping chromosomes, even in good quality metaphases, this should significantly simplify and speed up an automated process.

Dicentric aberrations vary considerably in appearance, both in length and positions of the centromeres (figure 1). If an automated system is programmed to attempt to detect all dicentrics, this may considerably increase the number of false positives that will need to be reviewed and eliminated by the operator. A sensible strategy may be to consider just those candidate aberrations that conform to a certain strict definition that encompasses only the more easily identified dicentrics (Piper *et al*, 1988). For example, dicentrics tend to be larger objects than the medium sized chromosomes, so acceptance could be limited to those objects that are not smaller than the C group (6-12). Identification of dicentrics is also more certain if they are "metacentric", ie. both centromeres being located away from the tips of the aberrant chromosome. Exclusion of those candidates with one or both centromeres being "acrocentric" will obviously result in a significant number of dicentrics being ignored, but again will lead to far fewer ambiguous images and false positives being referred to the operator.

The relative numbers of dicentrics that are meta- or acrocentric has never been measured, but gathering such basic data from a large enough sample using a conventional microscope should present no real problem. Having established how many, say 50%, of aberrations are on average deliberately lost by excluding the acrocentrics, the statistical accuracy of data could be restored by doubling the number of cells analysed. For such a strategy to be economically viable in terms of saved operator review time, it must of course more than halve the number of objects per cell referred for review, and in addition one would hope that the number of inadvertent false negatives is substantially reduced.

Given the requirement in mutation studies for perhaps thousands of cells to be examined per specimen, analysis will have to be done on metaphases that are generally less than superb karyotypable quality. Lesser quality spreads are usually those with more touching or overlapping chromosomes and generally these composites will be larger than the medium sized chromosomes. When scoring with a conventional microscope, these objects rarely present a problem for rapid interpretation, although two chromosomes touching end to end can be misleading. In operator assisted karyotyping such unsegmented images require the operator to draw a split line between them with a light pen or mouse. A rapid aberration system however, cannot rely on such a procedure and so automatic decomposition of clusters will be essential.

## 3. Sister chromatid exchanges (SCE)

As well as the classical aberration studies there is probably an even larger potential market for mutagen studies involving automated analysis for SCE (figures 2 and 3). Here, having found a metaphase, one needs to go a small

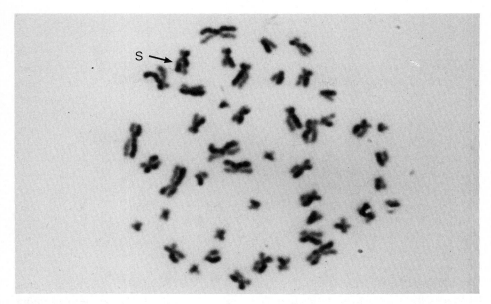

**Fig. 2.** A second division metaphase (M2) showing harlequin staining and containing a few SCEs.

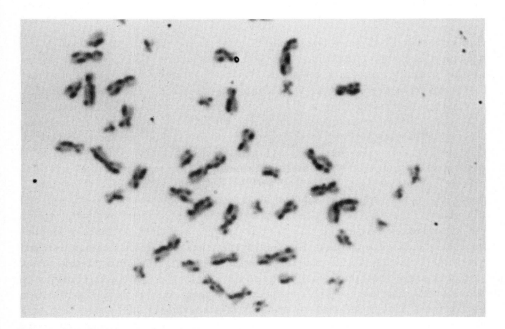

**Fig. 3.** A second division metaphase (M2) showing a large number of SCEs.

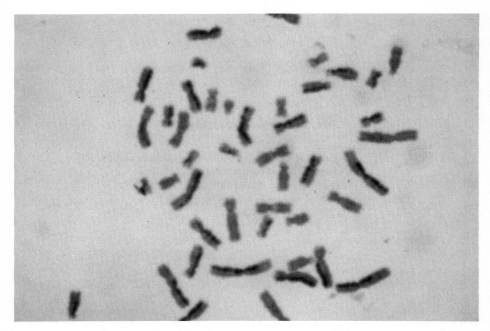

**Fig. 4.** A third division metaphase (M3) characterised by only 25% of the chromosome material staining darkly.

step further to distinguish the second division metaphases (M2) stained by the harlequin reaction. On a slide there will be a mixture of M1, M2, M3 (figure 4) and possibly even later divisions. With the original Cytoscan metaphase finder we have experimented with changing threshold values on the density acceptance levels and from a slide known by visual counting to contain 20% M2 cells, a list of metaphases comprising 60-70% M2 cells can be produced. This can probably be improved by proper program modifications directed to this specific end. Having found M2 metaphases, a major difference in SCE scoring when compared with "dicentric hunting" is in the number of aberrations present. Whereas dicentrics occur with a background frequency of perhaps 1 per 1000 metaphases, SCEs occur at a frequency of several events per cell (figure 2). This may be elevated after exposure to clastogens to several events per chromosome (figure 3). Thus it is usually sufficient to score a smaller number of cells per specimen. Most cytogeneticists wish to distinguish between centromeric switches and exchanges elsewhere along the chromosome arms, and it is likely that they would require an automated system to continue to provide this distinction. The specification of programs for SCE identification is a subject that I deferred to the software specialists. The presence of a simple reversal in the alternating light dark sequences along the length of the two sister chromatids identifies an exchange and this feature was used in a computer- aided analysis system described by Zack *et al* (1976).

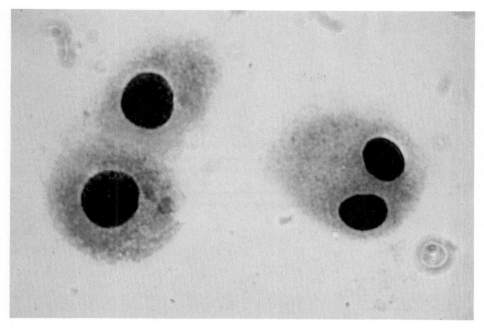

**Fig. 5.** A binucleate cytochalasin-blocked cell (right). An automated system will need to distinguish this from two touching mononuclear cells (left).

# 4. Micronuclei

Analysis for micronuclei in a variety of cell systems as an indicator of exposure to clastogens has been enthusiastically pursued for many years. Visually these images are far simpler and require less skills and time than the analysis of metaphases. The presence of a micronucleus leads to the possibility of selective death of a cell at division because loss of chromosome material reduces the likelihood of viable daughter cells being formed. Thus, quantification of data on micronucleus yield has sometimes been in doubt because of the need to incubate for lengths of time when it is known that there will be mixtures of cells in different *in vitro* cycles present on the slides. A recent important development with lymphocytes that overcomes this is to include cytochalasin B in the cultures (Fenech and Morley, 1985). This permits the nucleus to divide but stops the ensuing division of the cytoplasm, thus allowing micronuclei to be scored in binucleate guaranteed second *in vitro* interphase cells. This advance is somewhat analagous to the fluorescence plus Giemsa staining method which has allowed conventional aberration scoring to be confined to guaranteed M1 cells.

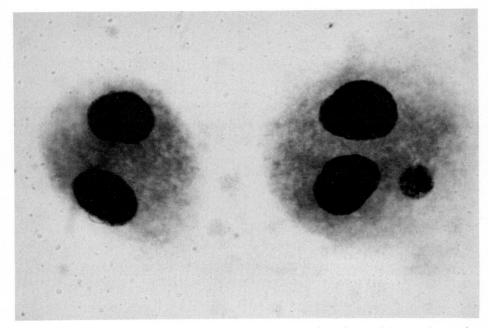

**Fig. 6.** Two cytochalasin-blocked binucleate cells with one (right) containing a micronucleus.

The images involved in micronucleus scoring seem highly amenable to automated analysis and indeed Callisen *et al* (1984) have described an automated
image acquisition, digitisation and analysis system. In contrast to metaphases
it is essential with micronucleus preparations that the cells retain their cytoplasm. This ensures that micronuclei are also retained and not lost during cell
fixation or slide making. For the cytochalasin method it is also important to be
able to see whether two large nuclei in close proximity are contained within the
same cytoplasm (figure 5). As clumping does occur, two touching mononuclear
cells with poorly preserved cytoplasm could be mistaken for a single entity.
The cytoplasm of the binucleate cell then needs to be searched for smaller cir-
cular inclusions, the micronuclei, that exhibit the same staining reaction as the
macronuclei (figure 6). Most cytogeneticists have traditionally used Giemsa
stain and this may not be ideal for micronucleus preparations. This is not
an exclusively nuclear stain and so the cytoplasm stains also. Whilst this is
an advantage, it means that other organelles in the cytoplasm are discernible
and may be mis-identified as micronuclei. Particularly for automated detection
of cytochalasin blocked cells and accurate discrimination of smaller objects in
their cytoplasm, specific staining regimes with good spectral separation for the
cytoplasm and nuclei plus micronuclei should be used.

Micronuclei are induced by a wide variety of clastogens so that their background levels are higher and more variable than the dicentric. Therefore it is unlikely that the faster acquisition of data that the micronucleus assay offers, especially if assisted by automation, can be used to lower the limit of detection (~0.3 Gy) of radiation dose for an individual. Its application lies rather in the ability to analyse quickly samples from a lot of people after a serious, Chernobyl-like, accident. By contrast automated dicentric scoring opens the prospect of significantly reducing the statistical uncertainties on individual low dose estimates in the region of 0.05-0.1 Gy.

# 5. Conclusion

In this paper I have tried briefly to review where we, the end users, stand, and inevitably the comments are influenced by the NRPB's experience to date which is confined to metaphase finding with a Cytoscan followed by visual analysis for classical aberrations. The considerable effort still needed to develop automation of scoring cells must be directed to those end points for which there is a commercially viable market. These are the classical aberrations, sister chromatid exchanges and micronuclei. I particularly want to draw to the attention of the systems developers the recent advance that has been made in the micronucleus technique and which is leading to a rapid and increasing enthusiasm for this assay among the biologists.

# References

1.   Callisen H, Norman A, Pincu M. Computer Scoring of Micronuclei in Human Lymphocytes. In: Eisert WG, Mendelsohn ML (eds) Biological Dosimetry, (Springer, Berlin, Heidelberg, New York 1984) pp.171-179
2.   Fenech M, Morley AA. Measurement of Micronuclei in Lymphocytes. Mutation Res. 147:29-36 (1985)
3.   Finnon P, Lloyd DC, Edwards AA. An Assessment of the Metaphase Finding Capability of the Cytoscan 110. Mutation Res. 164:101-108 (1986)
4.   Piper J, Towers S, Gordon J, Ireland J, McDougall D. Hypothesis Combination and Context Sensitive Classification for Chromosome Aberration Scoring. In: Gelsema ES, Kanal LN (eds) Pattern Recognition and Artificial Intelligence (Elsevier, Amsterdam, 1988) pp.449-460
5.   Zack GW, Spriet JA, Latt SA, Granlund GS, Young IT. Automatic Detection and Localisation of Sister Chromatid Exchanges. J. Histochem. Cytochem. 24:168-177 (1976)

# An Automated Chromosome Aberration Scoring System

Th. Lörch, C. Wittler, G. Stephan, J. Bille

## Summary

The statistical evaluation of structural chromosome aberrations is an accepted biological dosimetry method. This paper describes an automated dicentric chromosome detection system which is based on a real-time hardware preprocessor, a binary and a grey level image buffer, and a powerful multimicroprocessor and which has the potential to speed up aberration analysis significantly. The system finds metaphase cells on microscopic slides, segments metaphase images, and classifies single chromosomes as "normal" or "dicentric". A short description of the system hardware and the strategies and algorithms used, metaphase finder and segmentation performance data, chromosome classification results, and an execution time estimation are presented.

## 1. Introduction

The only biological dosimetry method now used routinely in radiation protection is the scoring of structural chromosome aberrations in human lymphocyte metaphase cells (figure 2) [1]. Especially the dicentric chromosome (figure 1) is a rather specific indicator of ionizing radiation and shows a low background frequency of about 1 in 90000 chromosomes [2]. The relationship between the radiation dose and the dicentric rate is given by a linear-quadratic curve. For statistical reasons, at low doses a large sample of metaphase cells must be analysed, which is a very time-consuming task.

**Fig. 1.** Normal (left) and dicentric chromosomes.

At the Institute for Radiation Hygiene of the Federal Health Office of the FRG, usually a sample size of 1500 cells is used, a complete analysis takes from

**Automation of Cytogenetics**  Editors: C. Lundsteen J. Piper·
© Springer-Verlag Berlin Heidelberg New York 1989

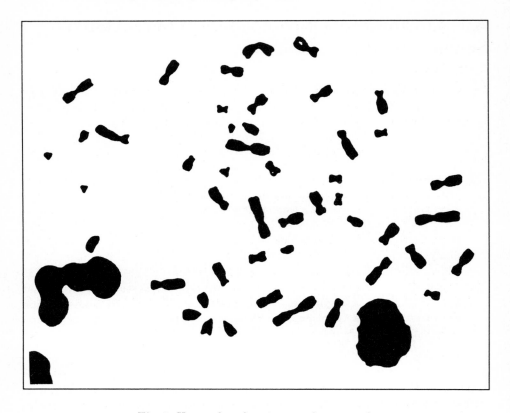

**Fig. 2.** Human lymphocyte metaphase spread.

5 to 10 days. The resulting high cost of a biological dose estimation restricts the application of this method now to few cases of suspected irradiation for which conventional dosimetry is either impossible or not sufficient.

One possible solution to this problem is the automation of the analysis by digital image processing. Due to the very low frequency of dicentrics at typical doses encountered in radiation protection (e.g. 1 dicentric in 18000 chromosomes for 0.05 Gy 250kV X-rays), a fully automatic scoring of dicentric chromosomes seems at present not to be feasible; this view is shared by other researchers [3,4]. Consequently, the goal of the system described in this paper is to find metaphase cells, to digitize and segment metaphase images, to identify candidate dicentrics, and to store their images and slide coordinates in unsupervised operation. In a final interactive step, all objects classified as "dicentric" are reviewed by the operator using a screen display or the microscope. To be efficient, the system has to be very fast during the automatic analysis and should not require too many interactions.

# 2. Materials and Methods

## 2.1. Hardware

Image generation is done by a conventional light microscope equipped with a plumbicon TV-camera and a motorized scanning stage. The microscope magnifications (from slide surface to plumbicon target) are 50x for metaphase search and 200x for aberration analysis. The stage has a capacity of 8 slides and can be moved by stepping motors with a resolution of $1\mu$m in X- and Y- and $0.05\mu$m in Z-direction. The maximum speed for the X- and Y-axis is 50 mm/sec (50000 steps per second). The video signal of the TV-camera is fed into three image processing devices:

(1) a hardware preprocessor which evaluates the signal in real-time and measures 5 parameters for every video line (object count, optical density, bandpass-filtered optical density, 1. and 2. gradient). Its data are suitable for fast preselection of frames during metaphase search and for automatic focussing.

(2) a binary image buffer used for a more detailed two-dimensional analysis of a frame.

(3) a grey level image buffer system with a 768x512 resolution and 64 grey levels (pixel spacing at 200x magnification : $0.075\mu$m). The aberration analysis is based on the digitized images generated by this system.

All computations are performed by a POLYP-polyprocessor-system [5]. It consists of 12 independently operating modules - each based on the Motorola 68000 CPU running at 10 MHz with 1 MByte of RAM - connected by a double 32-bit bus system, the POLYBUS [6]. The POLYP is controlled by a host computer which does all I/O operations, and it is connected to the Multibus-based devices mentioned above and to the stage control unit. The structure of the system hardware is outlined in figure 3.

## 2.2. Strategy

The performance of a classification system can be characterized by two error rates : the false positive rate (FP) and the false negative rate (FN). For dicentric recognition, FP gives the rate of non-dicentric objects identified as dicentrics, and FN is the proportion of all dicentric chromosomes not detected by the system. A constant FN rate can be fully compensated by the system's dose-response curve and by using a larger sample of cells to reach the same level of statistical significance. The cost of a FP decision is much higher, because every false positive must be examined in the interactive step. This causes an almost linear increase of interaction time with FP rate.

FN and FP rates are connected by a receiver operating characteristic (ROC) curve, i.e. for a given classifier one error rate can be selected and the other one is then defined by the ROC curve. This is illustrated in figure 4

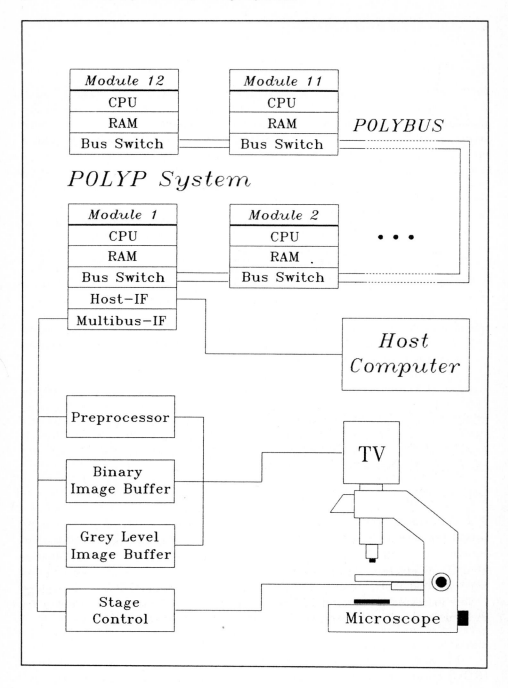

**Fig. 3.** Hardware structure.

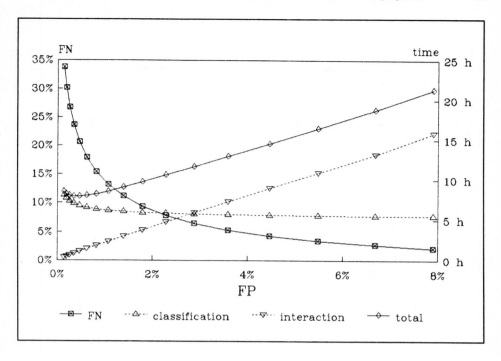

**Fig. 4.** Theoretical ROC curve and execution times.

showing a theoretical ROC curve for a linear classifier (FN versus FP) and the times needed for the automated classification, the interaction and the total analysis (based on the error rates and times given below). The total time has its minimum for FP = 0.5%, the corresponding values are FN = 20%, classification time 6.8h, interaction time 1.1h. This shows that for optimum overall performance the automated system should be tuned to a very low FP rate while a considerably higher FN rate can be accepted. Segmentation divides the image of the metaphase cell into subimages which ideally contain exactly one chromosome. In reality, objects lying close together cannot always be separated and result in subimages containing clusters of objects. These clusters should be detected automatically and can then either be decomposed by more sophisticated methods or just be ignored. We now use the second strategy : this increases the overall FN rate, but reduces the mean time needed for automatic classification and the complexity of the software.

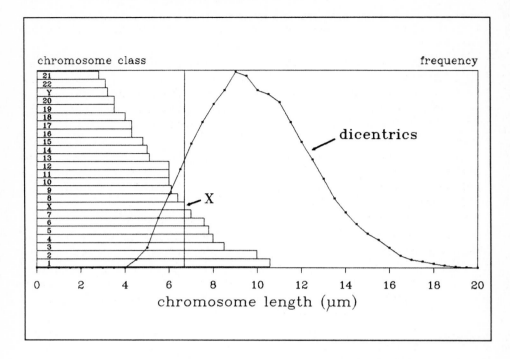

**Fig. 5.** Dicentric length distribution and normal chromosome lengths.

Dicentric chromosomes are produced by the cell's repair mechanism by connecting two chromosome fragments containing each one centromere. As centric fragments are on the average longer than acentric fragments, dicentrics tend to be longer than normal chromosomes. Figure 5 shows the theoretical length distribution of dicentric chromosomes resulting from a computer simulation [7] and the mean lengths of the normal chromosome classes [8]. Only 8% of the dicentrics are shorter than a X-chromosome, the corresponding value for normal chromosomes is 66%. This means that by ignoring all objects shorter than the X, the number of objects to analyse is reduced to one third, while the FN rate is increased only slightly. At the same time the FP rate is decreased significantly because small chromosomes have a much higher probability for false classification.

## 2.3. Algorithms

During metaphase search, the whole slide is scanned and each frame is evaluated by a three-step hierarchical classifier based on hardware preprocessor data and binary image features. Focussing is done by moving the stage up and down and maximizing a focus function. Suitable focus functions are the total above-threshold area or the 1. or 2. gradient of the video signal computed from the preprocessor data. A more detailed decription of metaphase search and focussing can be found in [9].

For segmentation of metaphase images to single chromosomes global grey level thresholding is used. Because for aberration scoring it is not necessary to restrict the analysis to complete cells (with 46 chromosomes), we consider at the moment the use of more sophisticated algorithms uneconomic. The segmentation threshold is computed from percentile values of the grey level histogram. The objects are then isolated by a fast contour following algorithm.

The first step of the chromosome evaluation is a rotation to an orientation for which the principal axis of the object coincides with the X-direction of the image matrix. The rotation angle is calculated from the second-order canonical central moments of the grey level distribution. This rotation makes it possible to compute chromosome profiles by simple and fast methods.

The classification is then performed in 5 steps. In the first step a box classifier using simple geometric parameters eliminates objects with a low a-priori probability of being dicentric (including especially small objects, see above). This fast preselection reduces the mean computation time per object considerably. The remaining 4 steps are based on linear statistical classifiers computed by multivariate discriminant analysis [10].

The goal of the 2. classification step is to recognize non-chromosome objects, e.g. undivided nuclei or chromosome clusters. Features used for this purpose are the mean deviation of the medial axis from a straight line, the variation of grey levels along the medial axis etc. By the classifiers 3, 4, and 5 the remaining objects are classified as "normal" or "dicentric". The 3. classifier eliminates all objects that are certainly not dicentric. In the 4. step distinct dicentric chromosomes are recognized. The 5. classifier is optimized to separate the remaining objects (which are the most difficult cases).

All features in the steps 3, 4, and 5 are extracted from various chromosome profiles. The basic profiles are the integrated density profile (IDP) and chromosome width profiles at two different grey levels. From each basic profile, derived profiles can be generated - e.g. gradient profiles or a profile's deviation from its convex hull. The latter two profiles are important because they are more "sensitive" than the basic profiles and therefore give better results for acrocentric centromeres (i.e. centromeres near the end of the chromosome, which often do not correspond to clear minima in other profiles).

For chromosome centromere regions, the integrated density profile and the width profiles normally show minima. Dicentric chromosomes have two centromeres and can be recognized by their profiles having two such minima instead of one in the case of normal chromosomes. To discriminate between these minima and profile minima caused by other phenomena (e.g. inhomogenous staining), the profiles are iteratively smoothed until two minima remain. For the less distinct of the two (which only for the dicentric chromosome corresponds to a centromere), the relative depth and relative width are computed

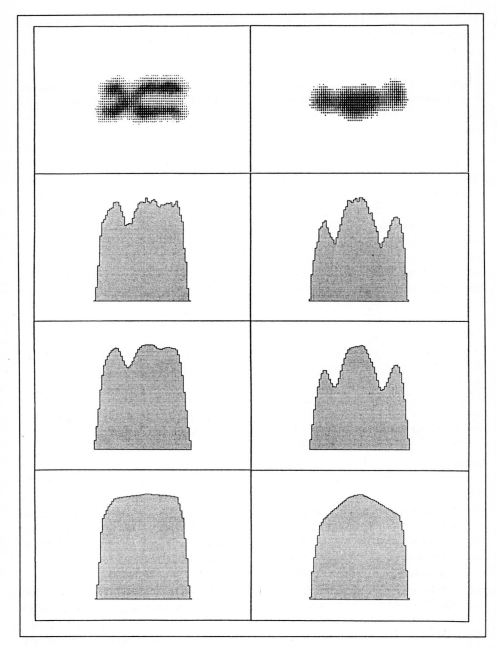

**Fig. 6.** Profile evaluation (left : normal chromosome, right : dicentric), from top : grey level image, integrated density profile IDP,IDP after smoothing, convex hull of IDP.

for all profile types and used as features for the classificators 3, 4, and 5. (Convex deviations differ from the other profile types insofar as they show maxima instead of minima at centromeres).

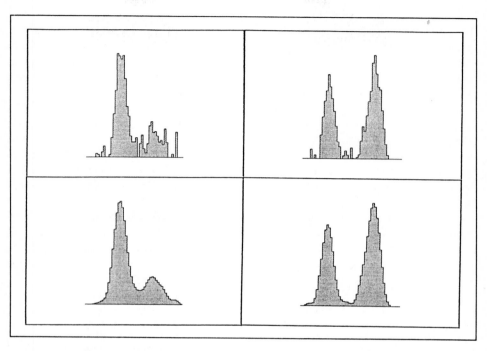

**Fig. 7.** Profile evaluation (left : normal chromosome, right : dicentric), from top : convex deviations of IDP, convex deviations after smoothing.

There are three important points to note concerning the classification algorithm :

(1) the hierarchic structure of the classification : the computationally costly higher levels of the analysis are executed only for relatively few objects for which all lower levels have yielded a positive result (i.e. "easy" objects are classified very quickly, for "difficult" objects the full analysis is carried out which gives low error rates).

(2) the interactive smoothing of the profiles : smoothing is iterated until two minima remain, thus adapting to inter-chromosome and inter-cell variations in profile contrast.

(3) the use of the less significant of the remaining profile minima for feature extraction : because a dicentric and a normal chromosome differ mainly in this minimum (and not in the more significant minimum which in both cases corresponds to a centromere) this improves the discrimi-

nation between the two classes substantially (compared for example to features based on the entire profiles).

## 3. Results

In a performance test based on a sample of 6 slides (unbanded giemsa-stained preparations of human lymphocytes), the metaphase finder gave a false positive rate of 1.8%, while 62% of all metaphases and 85% of all scorable metaphases were found. The search speed was 2.3 frames per second, on the average every 4 seconds a metaphase cell was found (for details see [9]).

As a test of automatic digitization and segmentation a data set of 13801 chromosomes was analysed for segmentation errors. After an automatic metaphase search, the cells were relocated, centered, focussed, digitized, and segmented without interaction at high magnification (200x). 1896 chromosomes were involved in clusters, and 80 chromosomes were fragmented by the segmentation algorithm. This means that about 86% of all chromosomes were segmented correctly.

Aberration analysis was developed and tested using a learning set of 7597 normal chromosomes, 98 dicentrics, and 1843 non-chromosome objects (nuclei, dust particles, segmentation errors etc.) from 13 slides. The following table shows the number of objects remaining after each classification step for all three classes :

|  | chromosome | dicentric | non-chr. |
|---|---|---|---|
| learning set | 7597 | 98 | 1843 |
| after preselection | 1912 | 94 | 125 |
| after 2. classification | 1731 | 92 | 28 |
| after 3. classification | 277 | 90 | 17 |
| after 4. & 5. classif. | 61 | 82 | 9 |
| final percentage | 0.8% | 83.7% | 0.5% |

This results in a total FP rate of 0.74% and a FN rate of 16.3%. (Note: these are error rates for the classification only. False negatives due to segmentation errors are not included). The time needed for all computations is 2.1 sec/object for one CPU, which lets us expect a time of 0.28 sec/object for the POLYP system. Based on an estimated time of 10 sec per decision for the interactive step and on the results given above, the following execution times for an analysis equivalent to 1500 cells can be calculated :

| metaphase search | 2270 | cells | 2.6 h |
|---|---|---|---|
| aberration analysis | 85500 | objects | 6.7 h |
| interactive step | 630 | objects | 1.8 h |
| total | | | 11.1 h |

Including the setup-procedures for the metaphase search and the aberration analysis, an operator must be present for about 2 hours - compared to between 5 and 10 days for a conventional analysis of the same size.

## 4. Discussion

The results obtained so far are encouraging and indicate that semi-automatic dicentric scoring is feasible. The development of the metaphase finder is completed, the programs for aberration analysis (written in PASCAL and 68000-assembler) are running on one CPU and are now being implemented in parallel on the POLYP system. Before the system can be used routinely, the significance of the dicentric rates found has to be established in a direct comparison with the conventional aberration analysis.

Obviously there exists a number of possible future improvements. The speed of the automated analysis can be increased by coding a greater proportion of the software in assembler language. The FN rate could be reduced by integrating an algorithm for chromosome cluster decomposition. Including additional features (computed e.g. from transverse profiles at centromere locations) or adding a classification step (for example testing hypotheses for the most frequent false positive causes) would decrease the FP rate.

*Acknowledgements.* The development of the dicentric scoring system was supported by the "Bundesministerium für Umwelt, Naturschutz und Reaktorsicherheit" of the Federal Republic of Germany.

## References

1.    Dolphin GW, Lloyd DC, Purrot RJ. Chromosome aberration analysis as a dosimetry technique in radiation protection. Health Physics 25:7-15 (1973)
2.    Stephan G. Chromosomenanalysen bei exponierten und vermutlich exponierten Personen, 1 Okt. 1982 bis 31 Dez. 1983. ISH Report 47, Inst. for Radiation Hygiene of the Federal Health Office, ISBN 3-924403-22-8 (1984)
3.    Piper J, Towers S, Gordon J, Ireland J, McDougall D. Hypothesis combination and context sensitive classification for chromosome aberration scoring. In: Gelsema ES, Kanal LN (eds) Pattern Recognition and Artificial Intelligence (Elsevier, Amsterdam, 1988) pp.449-460

4.  Mason D, Rutovitz D. The economics of automated aberration scoring. In: Evans HJ, Lloyd DC (eds) Mutagen induced chromosome aberrations, (Edinburgh University Press, 1978) pp.339-345
5.  Männer R, Deluigi B. A fault-tolerant multiprocessor without bottlenecks. IEEE Trans. Nucl. Sci. 28:390-394 (1981)
6.  Männer R et al. The POLYBUS : a flexible and fault-tolerant multiprocessor interconnection. Interfaces in Computing 2, pp. 45 (1984)
7.  Lörch T. Ein System zur automatischen Detektion dizentrischer Chromosomen. PhD thesis, University of Heidelberg (1986)
8.  Lundsteen C, Philip J, Granum E. Quantitative analysis of 6985 digitized trypsin G-banded human metaphase chromosomes. Clinical Genetics 18:355 (1980).
9.  Lörch T, Bille J, Frieben M, Stephan G. An automated biological dosimetry system. SPIE Proc., vol. 596, pp. 199-206 (1985)
10. Cooley WW, Lohnes PR. Multivariate Data Analysis. (John Wiley, 1971).

# Automatic High Resolution Digitisation of Metaphase Cells for Aberration Scoring and Karyotyping

Margaret Stark, Steve Farrow, Mark McKie, Denis Rutovitz

**Summary**

In this paper we describe techniques for automatic digitisation of a sequence of metaphase cells, using the MRC Human Genetics Unit's "Fast Interval Processor" cell finding and karyotyping machine. Results of experiments are presented, which demonstrate that relocation from metaphase search, auto-focus and centering are more than adequate for routine use of the system for aberration scoring and other cytogenetics tasks.

## 1. Introduction

The degree of automation required in systems for determining the genetic karyotype differs markedly from that needed for automatic aberration scoring. The reason is that, in karyotyping, a full analysis of a banded cell would take a skilled operator in the vicinity of 5 minutes, and if in addition a rearranged photographic karyogram is produced, another 15 to 20 minutes will be required. A semi-automated system, which still requires input from an operator for correction of segmentation, adjustment of centromeres and correction of final classification is therefore competitive if the total time, or at least the total involvement of the operator in the process, can be reduced below about 5 minutes. Indeed, studies by Lundsteen *et al* (1987) have shown that without an automated system, the whole process of searching for suitable metaphase cells, counting 10 of them and analysing 4 takes a skilled technician in a working clinical laboratory a little more than 1 hour on average, but with a machine at today's level of automation this is reduced to about 35 minutes.

Now consider aberration scoring: In a study by Lloyd and Purrot (1981) it was shown that operators could find and score cells at rates of over 50 per hour on a reasonably well sustained basis. Therefore, if automation is to bring any gain in this area, operator involvement has to be reduced to well below one minute per cell. Lloyd also showed that with cells prelocated by a metaphase finder, the operator scoring rate could be brought closer to 100 per hour (Finnon *et al*, 1986). The implication is then that for further automation to be of any practical use, the total time must be less than 30 seconds or so per cell. In reality, however, something much more radical is required. In studies

**Automation of Cytogenetics**   Editors: C. Lundsteen J. Piper
© Springer-Verlag Berlin Heidelberg New York 1989

of environmental exposure one would very much like to be able to detect, say, a doubling of the background rate of aberrations found in unexposed populations. In most environments the background rate of aberrations from all causes results in an incidence of around 1 dicentric chromosome per 700 cells (if all aberrations are taken into account the rate is somewhat higher, but dicentrics are radiation-specific and are also an easier target for automated systems). To detect the difference between rates of 1:700 and 1:350 (which might be caused by an exposure of about $10^{-1}$Sv, or twice the internationally accepted occupational limit) at the 95% confidence level we would have to score some 5,600 cells. To accomplish such a task on a significant number of cases, the need for a human operator has to be virtually eliminated.

**Table 1.** Stages in finding dicentric chromosomes

| Operation | Interaction Acceptable? |
|---|---|
| select 20× objective | yes |
| load slides | yes |
| find metaphases | no |
| remove slides, oil, reload | yes |
| select 100× objective | yes |
| select cells for analysis | no |
| step from cell to cell | no |
| centre and focus | no |
| digitise | no |
| check digitisation quality | no |
| correct segmentation | no |
| find centromeres | no |
| select dicentrics | no |
| review dicentric candidates | yes |

Table 1 lists the operations necessary for an automatic screening system for dicentric chromosome aberrations which includes metaphase finding, high resolution digitisation and analysis, and indicates the acceptability of human interaction at the various stages. It is probably not necessary to arrange for automatic objective change or slide oiling, albeit at the cost of some residual inconvenience. However, it would certainly be impracticable to use an operator to check on the quality of the cells selected by a metaphase finder, or to use an operator to assess and correct the segmentation, look at centromeres and so on. The only exception is in relation to reviewing candidate aberrations found by the system. Provided these are sufficiently few in number (i.e., that the false positive figure is low) the amount of operator involvement required for assistance in the final acceptance of aberrations is relatively modest (Lörch et al, 1989, Piper et al, 1988).

The Fast Interval Processor (FIP) is the MRC's enhanced prototype version of the Cytoscan range of machines made by Image Recognition Systems. FIP's hardware and software have been presented in detail elsewhere (Shippey *et al*, 1981, Lloyd *et al*, 1987); therefore this paper will describe only the salient features involved in "hands-off" digitisation. Correcting segmentation and subsequent operations as outlined in table 1 are described by Ji (1989a, 1989b) and by Piper *et al* (1988) and will not be discussed here.

# 2. System hardware description

## 2.1. Scanner

This consists of a Zeiss IM35 inverted microscope with a motorised stage capable of holding 5 slides, though thus far we have not made use of the multi-slide capability. The stage speed can be up to 20mm/sec for positioning and 7.3mm/sec during metaphase finding. Scanning for high resolution digitisation is much slower at about 50 $\mu$m/sec. There are three 1024 element linear charge-coupled device (CCD) arrays, one used for image capture and the other two for auto-focus during metaphase finding.

By using different magnifications and changing the size of step, the spatial resolution of the instrument can be varied. For metaphase search a 20× objective is normally used, which with 1$\mu$m step size gives a pixel spacing of 1.0$\mu$m × 0.625$\mu$m; for high resolution digitisation a 100× objective with 0.1$\mu$m step size, to give a pixel spacing of 0.1$\mu$m × 0.125$\mu$m.

The scanner measures optical density rather than intensity. Its resolution can be set to any one of four scales, each with 256 steps, the resolution of each step being between 0.01 OD and 0.0025 OD depending on the scale chosen.

## 2.2. Pre-processor Unit

This consists of purpose built electronics, which in fast search mode converts the signal from the scanner into above threshold "intervals", or segments of a scan line (Shippey *et al*, 1981). It compresses the data and extracts relevant features for output to the computer, for fast metaphase finding. For high resolution scanning another mode is used in which either all pixel values can be transmitted, or a background-cut is applied, which run-codes the pixels below threshold.

## 2.3. Computers

A Plessey MIPROC computer controls the scanner and runs real-time image processing software (Shippey *et al*, 1981), under the direction of a multiprocessing executive. In addition to the MIPROC computer a multiple Motorola M68000 and VMEbus-based configuration is used running under the

OS9 operating system. This part of the system is used for the analysis and classification of cells digitised at high resolution, and to provide a user friendly interface to the system.

# 3. Low Resolution Scan

## 3.1. Set-up

The initial set-up procedure for a scan requires the operator to perform a number of tasks, namely:

(a) Load the slide, enter identification
(b) Select 20× objective, check all microscope settings.
(c) Instruct the machine as to the area of the slide to scanned, and/or the number of metaphases to be found.

These take about 30 sec of the operator's time, after which no further interaction is required until the set-up of the high resolution scan.

## 3.2. High Speed Metaphase Search

Before the metaphase search begins two short preliminary scans are performed. The first "shading correction" scan is used to correct for differences in the output levels from different regions of the image sensor. This can be caused by either inequalities in the sensitivity of the individual photo sites or by uneven illumination of the field covered by the sensor. The second scan is done over a small, hopefully representative, area of the slide to construct a histogram of density values. The histogram is used to calculate a threshold value and an optical density range to compensate for differences between pale- and darkly-stained slides.

The instrument next performs the metaphase search. It scans the slide with a 500 $\mu$m wide swathe usually at 6000 steps/sec. moving the stage in 1 $\mu$m steps. The slide is kept in focus by a continually running focus process which compares the outputs of the two focus sensors (Shippey *et al*, 1981, Lloyd *et al*, 1987). The scan progresses until the whole area has been traversed or until the number of metaphases requested has been found. For each metaphase located, a record is made of the coordinates and feature values measured. At the end of the scan the metaphases are ranked by a quality measure based on a linear combination of feature values (Shippey *et al*, 1986). Depending on the area scanned and the stage speed the whole procedure takes up to 3 or 4 minutes.

# 4. High Resolution Scan

Figure 1 shows a flow diagram of the stages in the digitisation cycle.

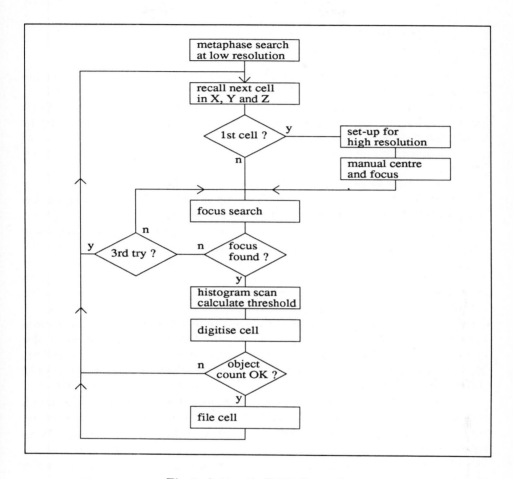

**Fig. 1.** Automatic digitisation cycle

## 4.1. Set-up

The first cell from the metaphase list is relocated in X, Y and Z. The operator is asked to change and oil the objective, adjust the optics, centre and focus the cell. The differences between the new operator selected position and the position stored in the metaphase list is used to calculate offsets in X, Y and Z, which are added to coordinates of all subsequent cells to compensate for

differences in position due to reloading and to the change-over from the 20×
to the 100× objective. This is the last manual task required until the complete
list of metaphases has been digitised.

Because the objective has been changed a second shading correction is
carried out.

## 4.2. Automatic Focussing

All metaphases after the first are automatically recalled at 100× at the
focus level at which they were encountered during high speed search at 20×,
corrected by the previously found offset. This procedure can give only a rough
focus position, to within 1μm or so, and a rapid optimisation is needed to
0.1μm, which is both the depth of focus at 100× and the nomimal resolution of
the focus motor motion. To do this a search is carried out around the recalled
position.

The contrast in the image at each level is available from the dedicated real-
time hardware which is used for maintaining focus during metaphase finding
(Shippey *et al*, 1981, Lloyd *et al*, 1987). The measure calculated is a function
of the difference between the optical densities of adjacent pixels, taken along
50 scan lines which cover a swathe 5μm wide that is centered on the recalled
cell position. This measure peaks for best focus and it has been found that
over a sufficiently large range of focus height it has the shape of the Gaussian
error curve. The implication of having a mathematical model for the variation
is that it is not necessary to evaluate the focus at every level within the search
range. Instead, a limited number of measurements permit interpolation to find
the peak, provided they bracket it and are well-behaved. Figure 2 shows a plot
of the contrast measure against focus height at increments of 0.1μm and the
best fit parabola through the points whose value is greater than one eighth of
the maximum.

A two-level strategy is employed. A rapid coarse search is made, which
in the majority of cases successfully finds focus. If it does not, a more time-
consuming sequence occurs. Initially a search is made over up to sixteen focus
levels separated by 0.8μm, which allows focus capture if true focus lies within
about 5μm of the recall level and the conditions are favourable. This takes
between 2.5 and 3.5 seconds depending on how far the search has to go to pass
the peak. To find the position of best focus, to within one step, when the
measurements were made at 8 step intervals, a curve is fitted to those points
which are not deemed to be outliers. The peak position is estimated, and a
move made to that level prior to digitisation.

This first fairly rapid, and usually successful, search can sometimes be
unsatisfactory because some condition on the contrast measures has not been
met. The program has only a very limited "view" of the metaphase compared
to an operator focussing on it and so can be misled. The criteria which have
to be satisfied are as follows. The step by step search is carried out from about

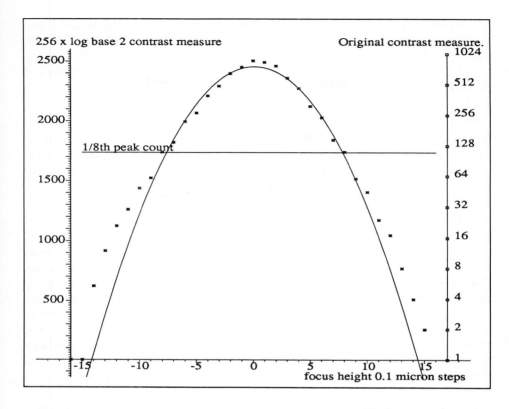

**Fig. 2.** Contrast measures obtained when scanning a metaphase with a 100× objective.

5μm below the recalled level to up to the same distance above. The search is only considered successfully terminated if the contrast measure has fallen to below one eighth of that at a previous peak value, provided that value was itself significantly large. An assumption is made, justified by the practical result, that measures greater than one eighth the peak value will fall sufficiently closely to the Gaussian error curve model to predict accurately the peak position, while those which are smaller are un-reliable. Curve fitting is only attempted if there is at least one accepted measure before the peak, and one after.

If the first search is unsuccessful the program can take "intelligent" action. Several situations and their probable corrections can be identified. If the peak occurred too close to one end of the search to allow interpolation, the search region can be moved up or down as appropriate and another attempt made. If the measures at the levels next to the peak are not large enough (because the width of the Gaussian is too small) then the separation between the levels can be reduced from 0.8μm to 0.4μm. If the peak contrast is too small, or the variation too little, then it is possible that either the recalled focus level was too far out for capture, or the $5\mu$m long swathe which was scanned did not contain enough material (or perhaps was dominated by a nucleus). In these cases a trial shift in Z, X or Y may bring success. Clearly attempts at correction could go on indefinitely: in practice only three are made, after which the cell is abandoned.

## 4.3. Threshold Calculation

When focus is found the cell is scanned, and pixel density values accumulated into a histogram. The program determines a suitable threshold value for the cell from the shape of the histogram. In order to minimise the time taken this scan and all positioning moves are done at high speed and $1\mu$m step size.

## 4.4. Digitisation

Unlike a TV camera based system, the size of the digitized area is not fixed. There is no size restriction in the X direction, whereas in the Y the size of the array limits the width to $128\mu$m. Using the dimensions of the metaphase found from the low resolution scan a suitable scanning window is selected. Large or irregular metaphase spreads only pose a problem if their span exceeds $128\mu$m in the Y direction, which should never occur in material used for aberration scoring (only very well spread prometaphase or prophase cells are likely to be so wide). Before the digitisation scan, the step size is changed from $1\mu$m to $0.1\mu$m. Using pixel-mode and in real-time, the data from the array is reduced by run-coding all pixels below the threshold value to permit rapid image processing.

## 4.5. Quality Control

Currently quality control is limited to a simple test which rejects the cell if the count of connected above threshold regions is outwith a preset range. This removes at the lower end very "tight" metaphases where most of the chromosomes are touching. Too high or too low a count could be due to poor thresholding. In either case dicentrics cannot be detected.

## 4.6. Digitisation cycle

Except for the first cell, all cells are relocated automatically using the X, Y and Z offsets obtained from the first cell. The operations of auto-focus, threshold calculation, digitisation and quality control are repeated on all cells until the metaphase list is exhausted.

Segmentation errors in accepted cells are automatically corrected (Ji, 1989a, 1989b), and the cells are analysed to identify any dicentrics (Piper *et al*, 1988). Further quality control decisions may be taken as the result of these analysis stages.

**Table 2.** The time taken for the operations in the digitisation cycle

|                                | time (sec) |
| ------------------------------ | ---------- |
| move cell - cell               | 3-4        |
| focus                          | 2.5-3.5    |
| threshold                      | 2          |
| digitise run-code and transmit | 9          |
| total                          | 18         |

# 5. Evaluation

## 5.1. Timing

A breakdown of the time taken for the different operations is given in Table 2, measured and averaged on 25 cells.

## 5.2. Centering and focus

The specimens used in the two following experiments were orcein stained metaphase preparations from blood lymphotyes irradiated with 2.5Gy X-rays. The object of the first experiment was to determine the success rate of automatic centering and focusing of metaphases at high power, the cells having been initially logged by the metaphase finder using a 20× objective. The slide was then removed from the instrument, oiled and replaced, and a 100× objective selected. The first recalled cell was centered and focussed. This experiment (Table 3a) showed that the original value for maximum allowable focus correction after relocation of 2.5$\mu$m was not sufficient.

In a second experiment the metaphases had been ranked in order of their likelihood of being scorable. Automatic digitisation was started at two points in the ranked list, at the "best" cell in the list, and at a middle ranked cell. It was found that in the 27 top ranked cells, centering was exact in 18 cases, but in any case the cell was always within the first-selected digitisation window. The autofocus procedure always found good focus, but in one case going to a 7.5$\mu$m excursion. All the cells bar one were complete, the exceptional one missing one chromosome. In general quality, 17 were judged "good" and the other 10 acceptable (Table 3b).

Using 19 middle rank cells, centering was mainly spot-on or at least within the digitisation window. The focus procedure always succeeded, though the

**Table 3.** Results of three experiments on automatic digitisation

| | | a | b | c |
|---|---|---|---|---|
| Metaphases: | found | 99 | 143 | 143 |
| | used | all 99 | top-ranked 27 | middle-ranked 19 |
| Centering: | bullseye | 34 | 18 | 14 |
| | within window | 64 | 9 | 3 |
| | lost | 1 | 0 | 2 |
| Focus: | pull-in range | $\pm 2.5\mu$m | $\pm 7.5\mu$m | $\pm 7.5\mu$m |
| | good | 79 | 27 | 19 |
| | poor | 20 | 0 | 0 |
| Cells complete/ | complete | 83 | 26 | 17 |
| broken: | missing 1-5 chr. | 16 | 1 | 2 |
| Metaphase | good | 45 | 17 | 10 |
| quality: | fair | 37 | 10 | 7 |
| | unusable | 17 | 0 | 2 |
| | garbage | 0 | 0 | 0 |

proportion of cells in which the excursion went to either $5\mu$m or $7.5\mu$m was greater. All cells were complete. As to quality, 10 were judged good, 7 acceptable, but 2 were unusable (Table 3c).

## 5.3. Quality of Digitisations for Karyotyping

Although aberration scoring is a prime objective of the work on automatic digitisation, we also examined the possibility of reducing by this means the amount of operator involvement in karyotyping. The most important question in this case is whether the quality of the digitisations of banded cells would be adequate for karyotype analysis. A pilot study was carried out using 5 amniocentesis slides, which were prepared by creating colonies on the cover-slips from small amounts of cell suspension and G-banded with trypsin. On each slide the highest ranked 42 metaphases were digitised and filed to disc. To test the quality of the cells and their digitisations as seen on the instrument's display, a cytogeneticist was asked to determine how many cells in the list needed to be examined in order to find 4 cells of quality adequate for karyotype analysis and 16 usable for counting. These results are presented in Table 4 which also shows for each slide the number of candidate metaphases from which the top 42 were taken. The object reject column shows how many digitisations were rejected automatically by the object count test. Table 5 shows a breakdown of reasons for rejection by the cytogeneticist. This shows that only about half of the rejections could be attributed to failures in the high resolution digitisation

**Table 4.** Automatic digitisation of amniocentesis colonies

| Slide No. | Clusters Found | Cells Digitised | Focus Rejects | Object Count Rejects | Cells Reviewed to obtain 20 |
|-----------|----------------|-----------------|---------------|----------------------|-----------------------------|
| BT6  | 339 | 42 | 0 | 4 | 25 |
| BT8  | 146 | 42 | 0 | 6 | 26 |
| BT10 | 192 | 42 | 0 | 6 | 25 |
| BT11 | 118 | 42 | 5 | 7 | 29 |
| BT13 | 170 | 42 | 0 | 5 | 27 |
|      |     |    |   |   | 132 |

**Table 5.** Cytogeneticists's criteria for rejection

| Reason for rejection | Number rejected |
|----------------------|-----------------|
| Too few chromosomes, cell maybe cut-off | 9 |
| Too many overlaps | 7 |
| Too pale | 7 |
| Focus Fuzzy | 4 |
| Incorrect Threshold | 5 |
|  | 32 |

cycle, the remainder being due to disagreement between the metaphase finder's and the cytogeneticist's rankings of the metaphases. The important point is that of the high ranked cells, three quarters were both successfully digitised and of sufficient quality for analysis.

# 6. Discussion

We were encouraged by the results from the experiments. The fine focus search has since proved invaluable even in a semi-automatic system, as the program finds the best focus for the digitising array, which is not always the same as for the eye of the observer, and auto-focus is now used routinely for all digitisation. The centering of cells in the array was adequate but could be improved by use of a more intelligent program. Such a program would be expected to recognise an incomplete cell, then re-adjust the stage position or window, to ensure digitisation of the whole cell. Another improvement which could be made would be to have feed-back from an evaluation of the result of digitisation to cause another attempt at focussing if it appeared to be poor.

Such methods might improve the yield of quality cells but increase the overall time taken as two scans would be required in all doubtful cases. Work is in progress to find a more reliable threshold algorithm as the present one cannot be used with confidence in all situations. The problems we thought we might encounter such as vibration caused by rapid and frequent stage movement have not arisen.

The automatic system has been used to collect data for an aberration scoring project and to date this data base contains 600 cells. Smaller data bases have also been collected for many different applications e.g. mouse, chinese hamster, fragile X and *in situ* hybridisation. Such was the consistency of results on the banded slides that all digitisations are now done without the help of an operator. Using this level of automation for karyotype analysis of banded cells the system has the potential to run "hands-off" until the karyotype correction stage.

In the karyotyping system developed by the Medical Research Council, now being marketed by Image Recognition Systems in their Cytoscan machine, the time for analysis of a cell is a function of the number of processors provided in the system. This system is described elsewhere (Piper and Rutovitz, 1986, Lloyd *et al*, 1987); it suffices here to say that it is a loosely coupled parallel processing system based on Motorola MC68000 family processors on a VME bus. The system is optimised for dealing with large processing components; typically one processor analyses one segmented object at a time. Applied to chromosome analysis problems, speed is an almost linear function of the number of processors. A parallel processing system will allow us to overlap analysis of the previously digitised cell with the digitisation of the next.

We began by stating that if in an aberration scoring context we could reach times of less than 30 sec per cell we would be competing favourably with the manual system. The times shown in table 2 are well within this limit. However, of greater significance is the fact that the machine is running unattended, reducing operator involvement to zero.

*Acknowledgements*. The authors would like to thank Judy Fantes for providing the orcein stained slides and giving her expert advice on the quality of the banded cells. The authors also wish to thank Jim Piper for many useful discussions and Jill Gordon for her endurance and useful feed-back while the system was in the development stage.

# References

1.    Finnon P, Lloyd DC, Edwards A. An assessment of the metaphase finding capability of the Cytoscan 110. Mutation Res. 164:101-108 (1986)
2.    Lloyd DC, Purrot RJ. Chromosome Aberration Analysis in Radiological Protection Dosimetry, Radiation Protection Dosimetry 1:19-28 (1981)

3.  Lloyd D, Piper J, Rutovitz D, Shippey G. A Multiprocessing Interval Processor for Automated Cytogenetics. Applied Optics 26:3356-3366 (1987)
4.  Ji L. Decomposition of Overlapping Chromosomes. This volume (1989a)
5.  Ji L. Intelligent Splitting in the Chromosome Domain. Patt. Recognition (in press, 1989b)
6.  Lörch T, Wittler C, Stephan E. An Automated Chromosome Aberration Scoring System. This volume (1989)
7.  Lundsteen C, Gerdes T, Maahr J, Philip J. Clinical performance of a system for semi-automated chromosome analysis. Am. J. Hum. Genet. 41:493-502 (1987)
8.  Piper J, Rutovitz D. A parallel processor implementation of a chromosome analysis system, Patt. Rec. Letts. 4:397-404 (1986)
9.  Piper J, Towers S, Gordon J, Ireland J, McDougall D. Hypothesis combination and context sensitive classification for chromosome aberration scoring. In: Gelsema ES, Kanal LN (eds) Pattern Recognition and Artificial Intelligence (Elsevier, Amsterdam, 1988) pp.449-460
10. Shippey G, Bayley R, Farrow S, Rutovitz D, Tucker J. A Fast Interval Processor. Patt. Recognition 14:345-356 (1981)
11. Shippey G, Carothers A, Gordon J. Operation and performance of an automatic metaphase finder based on the MRC Fast Interval Processor. J. Histochem. Cytochem. 34:1245-1252 (1986)

# Part II

# Karyotyping Systems: Progress and Evaluation

# Athena: A Macintosh-Based Interactive Karyotyping System

Lucas J. van Vliet, Ian T. Young, Ton K. ten Kate, Brian H. Mayall,
Frans C.A. Groen, Roelof Roos

## Abstract

In this article we describe a system we have constructed that provides for automated karyotyping of metaphase spreads. The software system **Athena** – based upon the Macintosh II computer and a Data Translation's frame grabber – is written entirely in C and consists of approximately 200 Kbytes of executable code. While **Athena** does not provide (at this time) facilities for automated metaphase finding, it does provide facilities for automated image segmentation into individual chromosomes, automated measurements on each banded chromosome, and automated classification into the standard Paris-convention karyotype. Further, the system provides the ability to construct one or more chromosome data bases to represent the types of metaphase spreads and staining techniques that may be used in a given laboratory.

Because we believe that it is impossible to construct a system that can achieve perfect segmentation, perfect separation of touching and overlapping chromosomes, perfect localization of the centromeres, and perfect classification, the system offers the possibility for interaction at each of the above stages using the well-accepted Macintosh user interface.

The processing time for the complete analysis of a $512^2$ image that contains a metaphase spread, interphase nuclei, debris and so forth depends heavily on the 'quality' of the metaphase image. Depending upon the number of interactions required to complete the segmentation, the total processing time varies from less than 3 minutes for nice metaphase spreads to about 6 minutes when many chromosomes need to be separated manually using the mouse. Of that time 90 seconds represents the automated processing and the rest the time for human interaction. This is on a personal computer that uses a MC-68020 processor with a clock frequency of 16 MHz and a MC-68881 floating point co-processor.

**Automation of Cytogenetics**   Editors: C. Lundsteen  J. Piper
© Springer-Verlag Berlin Heidelberg New York 1989

# 1. Introduction

For more than 25 years medical scientists, physical scientists, mathematicians, engineers, and computer scientists have attempted to automate the analysis of metaphase chromosome spreads. In an important sense automated karyotyping was one of the first problems in digital image analysis and pictorial pattern recognition. That there are still a number of on-going and healthy research efforts in this field – approximately 25 years after the publication of Ledley [1964, 1965] and Neurath et al [1965]– shows that the problem is both exquisitely difficult and extremely important. It is important to remember, however, that it was less than 35 years ago that the correct number of chromosomes in the normal human chromosome complement, 46, was first enumerated [Tjio and Levan, 1956]. Further, the accurate and reproducible identification of each of the 24 possible chromosome classes by cytogenetic experts was only possible after the development of the banding stains by Caspersson and his co-workers [1970a, 1970b].

Thirty years ago digital image processing was in its infancy. Computers were located in inaccessible computation centers. Frame grabbers did not exist. Computer languages were limited to assembly language. A few simple measurements were being performed [Schreiber, 1956; Deriugen 1957]. The first laboratory-oriented computer was introduced in the early 1960's with the PDP-1 and the first collection of languages appeared at about the same time. As of this writing (1988) the computational power, memory, and storage capacity available 30 years ago in computation centers has been substantially surpassed by the power available in a desk-top personal computer that is priced at a level that makes it affordable to individual researchers *and their families*. The software sophistication available on all fronts – operating systems, user interfaces, languages, and image processing algorithms – has witnessed a comparable improvement [Duin, 1983; Krusemark and Haralick, 1983; Preston, 1983; Tamura et al, 1983].

With respect to the problem of automated karyotyping, the implications of these improvements have become evident. As of this writing (1988) more than 10 companies are producing systems for the computer processing of digitized chromosome images. These systems range from those that can only be described as "electronic scissors" to systems that combine metaphase finding as well as pattern classification of banded chromosomes to achieve a result that requires a bare minimum of human interaction.

In this same context, three of us (Young, Groen, Mayall) have been involved for more than 15 years in research into various aspects of the quantitative and automated analysis of digitized chromosome images. This research has covered virtually all aspects of the quantitative analysis of chromosome images:

  • Metaphase finding [Bishop and Young, 1977]

- Automatic focussing [Mendelsohn and Mayall, 1972; Groen et al, 1985]
- Analysis of banding patterns [Granlund et al, 1976; Visser, 1981; ten Kate et al, 1983]
- Centromere location [Visser, 1981; de Muinck Keizer, 1984]
- Chromosome aberrations [Mayall at al, 1974; Zack et al, 1976; Mayall at al, 1977a; Mayall at al, 1977b]
- Accurate DNA measurements from digital images [Mendelsohn et al, 1969; Mendelsohn et al, 1973; Mendelsohn et al, 1974; Groen and van der Ploeg, 1978; Mayall et al, 1984]
- DNA species within chromosomes [Young et al, 1983].

On the basis of these developments we decided in 1983 that we were in a position to construct a karyotyping workstation. With the exception of automated metaphase finding we had assembled the components required for automatically producing a karyogram given a metaphase cell. These components consisted of:

1) image segmentation to find the individual objects;
2) procedures for eliminating non-chromosome-like objects;
3) procedures to separate touching chromosomes;
4) an accurate method for chromosome rotation;
5) an accurate procedure for centromere determination;
6) an intuitive but quantitative way to describe the banding patterns, and;
7) a context-sensitive classification procedure based upon the length, centromeric index, and the aforementioned band description.

These components – together with a window/menu user interface – were implemented (in C) as a complete system on a Vicom image processing system and formed the Master's thesis research of ten Kate[1985]. As a system the Vicom was appropriate for this development effort but as a long-term solution it was inappropriate. The system was too expensive, the operating system clumsy, the C compilers marginal, and floating point calculations excruciatingly slow. To remedy this situation we decided in the Spring of 1987 to "port" the karyotyping package to an Apple Macintosh II personal computer. This system offered an inexpensive, powerful platform for the chromosome project. Aside from the 16 MHz MC-68020 processor, high-speed hard disk, multi-megabyte RAM and MC-68881 floating point co-processor, the system also has an integrated 8-bit deep display and a professional software development environment, MPW [APDA, 1988] offering C, Pascal, object-oriented Pascal and assembly language.

The combination of our existing software, the Macintosh II platform and recently available frame grabbers has led to the development of **Athena**, a Macintosh-based interactive karyotyping system. The minimum Macintosh II configuration required for running the package consists of 4 Mbytes of RAM, a 40 Mbyte Hard Disk, a video expansion kit to provide 256 colors/grey levels on the screen, keyboard and mouse, and a color monitor. To provide hard copy, a Postscript-compatible laserprinter is necessary. In the remainder of this article we shall describe the software system as well as the preliminary results that have been achieved using the techniques contained in **Athena**.

## 2. Using Athena

**Athena** – much like a sister program **Acuity** developed for cell analysis on the Macintosh [Young and Roos, 1988] – is organized around the concept of an experiment. The user initiates a session with the program by "double-clicking" (or opening) the program icon shown in Figure 1. Thus the user initiates and interacts with the program along the lines defined in the Macintosh user protocol [Apple, 1985-1987].

**Figure 1**: Left - The icon for the **Athena** program. Right - The icon for an image

After starting the program the user is offered a number of menu choices as shown in Figure 2.

**Figure 2**: Left - The *File* menu that is used to define/open an experiment. Right - The *Goodies* menu that is used to preview an image, adjust the display contrast, or reformat image data.

The *File* menu – aside from several administrative functions – offers the user the ability to start a new experiment or to continue a previous one. The other possible choice after starting the program is the Goodies menu that allows the user to import, reformat, or display an image.

| | |
|---|---|
| **Untitled** | |

**Generic Image Name:** | chromosomes
**Number of Images:** | 1
**Image type:** ⦿ **Absorption**
    ○ **Fluorescent**
**Image Width:** | 512
**Image Height:** | 512
**Comment:**     Objects analyzed: 0

Segmentation Procedure
Analysis Procedure
Classification Procedure
Features for Classification
Time Stamp

Monday, March 7, 1988; 9:51:14 PM
Demonstration of Athena karyotyping program.

**Figure 3**: Experiment window associated with an – as yet – Untitled experiment. By clicking on the buttons *Segmentation, Analysis, Classification* and *Features*, the user can adjust the parameters associated with the various phases of the program.

After opening an old experiment (or starting a new one) the user is presented with a window as shown in Figure 3. By filling in the various entries in this window and the four sub-windows –*Segmentation, Analysis, Classification, Features* – the user defines the parameters, data sets, and so forth that are used when the program is executed. Arbitrary text as well as the current date and time can be entered in the comment block.

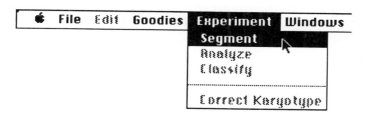

**Figure 4**: Menu choices associated with the various sequential steps in the karyotyping process.

At each of the three stages in the processing of the metaphase spread the user is presented with the opportunity to interact with the program. The three stages are intended to be used sequentially and are illustrated in Figure 4.

In the following sections we shall discuss each of these stages and the menu choices (as depicted in Figure 3) that are associated with them.

## 3. Segmentation

The fundamental technique used for segmentation is a combination of shading correction, thresholding, and binary image operations. Shading correction is offered as an option to remove (possible) effects associated with camera sensitivity or the illumination conditions. Shading correction is accomplished through the use of the grey-level morphologic filtering operations dilation (maximum) and erosion (minimum). The estimate of the background shading $S$ from an original image $I$ is given by:

$$S = Min_n( Max_n(I) ) \tag{1}$$

where $n$ is the neighborhood (support) of the filter. In this case the filter is always square of size $n$ and the value of $n$ should be chosen to be larger than the diameter of the largest object (usually an interphase nucleus) to be found in the image. The corrected image is then given by:

$$C = WHITE + ( I - S ) \tag{2}$$

where **WHITE** is simply the maximum possible value (usually 255) in the absorption image.

Thresholding is used to convert the image into a binary representation where the chromosomes and other objects are black (1) and the background is white (0). The various algorithms used to choose the threshold are displayed in the Segmentation window (Figure 5). The threshold selection algorithms will not be described here; the user is instead referred to Young and Roos [1988], Ridler and Calvard [1978], and Vossepoel et al [1979].

```
┌─────────────────────────────────────────────────────────────┐
│ ┌─────────────────────────────────────────────────────────┐ │
│ │         Segmentation Procedure Specification:           │ │
│ │ ☐ Shading Correction            ☒ Edit Mask             │ │
│ │    Shading Filter Size:  [31]                           │ │
│ │       Choose 'Interactive Threshold' or one of the Algorithms... │ │
│ │ ○ Interactive Threshold      ○ Minimum Between Two Maxima │ │
│ │ ○ ISODATA (Ridler & Calvard) ○ UniModal Background Symmetric │ │
│ │ ◉ Maximum Triangular Distance ○ Minimum After Maximum    │ │
│ │    Offset from peak:  [20] %  ○ Fixed Threshold Value: [128] │ │
│ │ ○ DIODA algorithm                                       │ │
│ │       Iterations Erosion Before Exo-Skeleton:  [2]      │ │
│ │     Iterations Erosions For Removing Small Filth: [3]    │ │
│ │   Iterations Erosions For Removing Large Objects: [10]   │ │
│ │                         [ Cancel ]  [ OK ]              │ │
│ └─────────────────────────────────────────────────────────┘ │
└─────────────────────────────────────────────────────────────┘
```

**Figure 5**: Choosing *Segmentation Procedure* offers the user the opportunity to specify the parameters associated with shading correction, thresholding, and binary image filtering. In addition, the user can indicate whether he/she wishes to correct interactively the segmentation result before the second stage – analysis – begins.

Binary image filtering is used to reduce the inevitable "false positives" and "false negatives" produced by thresholding. The operations *erosion*, *dilation*, *propagation*, *exclusive-or*, *skeletonizing*, and *and*ing are used to eliminate small artefacts and large objects, to fill holes inside chromosomes, and to separate touching chromosomes. The actual algorithms used to implement these operations are based upon the recently developed techniques of Van Vliet and Verwer [1988].

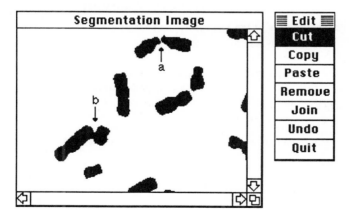

**Figure 6**: Near end of the segmentation phase, the user is offered the opportunity to correct errors if *Edit Mask* has been chosen (see Figure 5).

At the completion of the segmentation phase the user can correct the remaining errors through the use of the mouse and a menu interface (Figure 6).

We try to separate slightly touching chromosomes by a few erosions after which the background skeleton (exo-skeleton) is computed. This exo-skeleton forms dividing lines to separate the touching chromosomes. At point *a* in Figure 6, the exo-skeleton operation has successfully separated the two touching chromosomes. At point *b*, however, too few erosion passes (see Figure 5) were used to separate the chromosomes. By choosing *Cut* from the "pop-up" menu, the user can separate the two chromosomes. In the actual Macintosh display the user sees the grey-level image with contrast-stretching to maximize the visibility of details. The presumed chromosome borders are outlined in red. The mouse-based interactive procedure is straightforward. At the end of the segmentation phase each object – presumably a single chromosome – has been found, labeled and counted.

## 4. Analysis

The analysis phase consists of 1) rotating the individual chromosomes so that their long axis is parallel to the y-coordinate in the image [Groen et al, 1976], 2) determining the centromere position using the technique developed by Visser [1981] and 3) measuring the chromosome length, the centromeric index, and the band descriptors as described by Visser [1981]. The parameters associated with this phase are shown in Figure 7. This window is activated by selecting *Analysis Procedure* (see Figure 3).

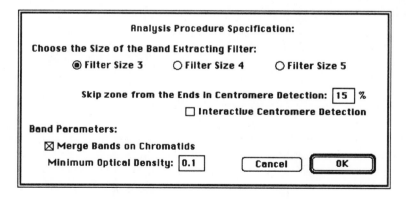

**Figure 7**: The parameter window associated with the analysis phase.

## 4.1. Measuring the centromeric index

We define the centromeric index as:

$$CI = \frac{\text{Length of the short arm (P terminal)}}{\text{Total length of chromosome}} \times 1000 \qquad (3)$$

This produces a number in the interval $0 < CI < 500$. Automatic centromere detection has proven to be a difficult task and none of the techniques available achieves 100% success. Many techniques are based upon a search for a pair of concavities along the chromosome contour. Pairs of opposite concavities then form candidates for the centromere position. **Athena**'s automatic centromere detector searches for two points having the shortest distance between the left and the right contour of a chromosome. In order to avoid detection at the ends of the chromosomes (telomeres), a certain distance from the top and bottom of the chromosomes is skipped. The user specifies this value as a percentage. The default value of 15% approaches the centromeric index of the acrocentric chromosomes of classes 13 through 15.

As of this writing the interactive correction of centromeric position has not yet been implemented. The interface will resemble that of "Fat-Bits" except with grey levels. That is, the user will be presented with an enlarged, grey-level image of the chromosomes that he/she chooses for centromeric editing. A "pop-up" menu such as that shown in Figure 6 will then guide the user.

## 4.2. Measuring the band parameters

Assuming absorption imagery, bands are considered as dark regions of the chromosome. An optical density-based threshold selects the dark parts as regions potentially bearing a band. In this way the detection of vague bands and vague connections between clearly separated bands is avoided. The default value of the threshold is set to 0.1 which corresponds to a transmittance of $\approx$80% in the band relative to the background.

The selected regions are then Laplace filtered along the main (medial) axis. The filter size (in Figure 7) refers to the size of the Laplace filter convolution window. This second order derivative filtering leads to the detection of hills (negative areas) and valleys (positive areas) in the grey value image. All hill-points are labeled and every set of connected points forms a candidate for a band. For each band, parameters are calculated such as: area, total optical density, begin, end, and middle position relative to the top. If "Merge..." is set, the system tries to match bands on separate chromatids based on their projected begin and end locations. From this information a subset of the bands is extracted and used for classification. **Athena** uses the central position of the following bands:

- Darkest band on the chromosome
- Band having the largest area
- First band on the p-terminal
- Last band on the q-terminal
- First band after the centromere position on the q-terminal.

After the rotation, the centromere localization, and the band measurement, the seven features – length, CI and the five band positions – are assembled (see Figure 8) and stored on disk in the same file that contains the experiment description as shown in Figure 3. The user may examine the data by summoning the *Features* window from the *Windows* menu.

| obj# | length | cIndex | firstB | lastB | LargeB | centB |
|------|--------|--------|--------|-------|--------|-------|
| 1 | 480 | 441 | 166 | 843 | 166 | 562 |
| 2 | 380 | 196 | 236 | 763 | 763 | 236 |
| 3 | 360 | 162 | 138 | 736 | 389 | 389 |
| 4 | 420 | 325 | 226 | 761 | 226 | 464 |
| 5 | 580 | 443 | 181 | 767 | 767 | 560 |
| 6 | 320 | 370 | 218 | 750 | 484 | 484 |
| 7 | 530 | 355 | 169 | 811 | 518 | 518 |
| 8 | 360 | 193 | 375 | 750 | 375 | 375 |
| 9 | 320 | 371 | 484 | 687 | 687 | 484 |
| 10 | 450 | 375 | 133 | 755 | 755 | 544 |
| 11 | 440 | 334 | 204 | 818 | 488 | 488 |
| 12 | 550 | 320 | 200 | 836 | 836 | 390 |
| 13 | 250 | 150 | 560 | 560 | 560 | 560 |
| 14 | 690 | 500 | 181 | 877 | 326 | 659 |
| 15 | 200 | 266 | 700 | 700 | 700 | 700 |

**demo – List Mode**

**Figure 8**: The measurements per chromosome are displayed and stored in a list mode. Object number refers to the order in which the chromosomes were found in the image and *not* to their class. The positions of the bands are relative to a normalized length of 1000 per chromosome. Thus object #1 has a last band at 84.3% of the distance from the top of the chromosome. Six of the seven measurements are shown here. The last measurement is seen by "scrolling" to the right.

While the features used here are intuitive and give satisfactory classification results (see Results below), it is a simple matter to substitute another set of feature measures such as the weighted density distribution measures of Granum [1981]. In a similar fashion other techniques can be used to determine the band positions such as the non-linear Laplace filters [Van Vliet et al, 1988] instead of the conventional Laplace filters.

## 5. Classification

The final phase of processing is the classification of the chromosomes on the basis of the measured features. Two different windows (as indicated in Figure 3) are associated with this task. The first window, *Classification Procedure* (see Figure 9), allows the user to indicate the name of the database that will serve as the standard for classification, the training set, the name of the database that *can* be built as a given metaphase is classified, and some parameters associated with the classification algorithms.

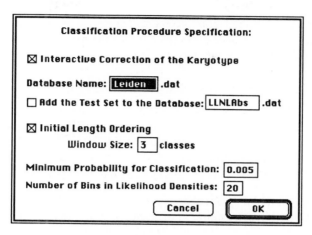

**Figure 9**: The first window, *Classification Procedure*, associated with the classification phase. The various parameters are discussed in the text.

Before classification we have measured the feature vector, **x**, for each unclassified chromosome. Further, we know the estimated frequency distribution of a feature vector for each of the 24 chromosome classes $p(x|\omega_j)$ (from the training database, e.g. Leiden.dat) and the *a priori* probability for a chromosome to belong to a class j, $P(\omega_j)$. The *a priori* probability is based upon biological information. (A healthy human being has 22 homologous pairs and 2 sex chromosomes). This information offers the opportunity to use the non-parametric Bayes' rule to calculate the *a posteriori* probability $P(\omega_j \mid x)$ that each individual chromosome belongs to class $\omega_j$:

$$P(\omega_j \mid x) = \frac{p(x \mid \omega_j)\, P(\omega_j)}{p(x)} \qquad (4a)$$

where

$$p(x) = \sum_{j=1}^{24} p(x \mid \omega_j)\, P(\omega_j) \qquad (4b)$$

This rule can be shown to produce, on average, the classification with the minimum number of errors.

For practical reasons we assume that the features are independent, so equation 4a can be rewritten as:

$$P(\omega_j \mid x) = \frac{\prod_{n=1}^{7} \alpha_n\, p(x_n \mid \omega_j)\, P(\omega_j)}{P(x)} \qquad (5)$$

where $x_n$ is an element of the feature vector $x$, that is, one of the seven measurements and $\{\alpha_n \mid n=1, \ldots, 7\}$ are weighting coefficients (see Figure 10.) For more information about the Bayes' rule see Duda and Hart [1973].

## 5.1. Estimating the frequency distributions

The frequency distributions (or class conditional probability density functions) are estimated from a learning set of patterns (database) with known class membership. Because the band features used are non-parametric, the problem of estimating $p(x \mid \omega_j)$ cannot be reduced to the problem of estimating the parameters of a distribution such as the mean and standard deviation of a hypothesized Gaussian.

**Athena** uses the histogram approach to transform a data set into a probability density function. The range of each feature $x_i$ of vector $x$ is divided into a fixed number of equal intervals or bins. The number of data points falling into each bin is counted and forms the basis for the probability estimate. The number of bins strongly depends on the size of the learning set and the underlying distribution. For Gaussian distributions the number of bins should be about $\sqrt{N}$ (if $N$ is the size of the learning set). Because the distribution of each chromosome's length is approximately Gaussian and fills about one third of the total range, $3\sqrt{N}$ is a good estimate for the total number of bins to use. As an example, if we have 40 chromosomes per class, then the number of bins should be about $3\sqrt{40} \approx 20$ (see Figure 9).

## 5.3. Context-sensitive classification

After computing the *a posteriori* probabilities for each chromosome a context sensitive classifier assigns the chromosomes to the possible classes. It is assumed that the metaphase contains two copies of each autosome and two sex chromosomes, xx for females and xy for males. An exception is made when there is an abnormal number of chromosomes in the cell, e.g. Down's syndrome. The classification is a two-step process.

In the first step, all chromosomes are ordered in decreasing length, building a list with the longest chromosomes first. The chromosome classes are also ordered in a list of decreasing length: 1 through 7, x, 8 through 22, and y. These two lists are then associated with each other. The first and second chromosomes from the length list with chromosome class 1, the next two chromosomes with chromosome class 2 and so forth.

After initial classification based upon length, we assign cost factors to the surrounding classes that make a transfer to another class over longer distances less likely. Distance is defined as the absolute difference in class number between start class and end class. Thus moving a chromosome from class 2 to class 6 is a distance of 4. **Athena** (in the current version) assigns no cost $(C = 0)$ to classes with a distance smaller than or equal to a pre-defined window size (see Figure 9) from the current class leaving the *a posteriori* probabilities $P(\omega_j \mid x)$ for j in **(class-window)** $\leq$ **j** $\leq$ **(class+window)** unchanged. An infinite cost $(C = \infty)$ is assigned to all classes with a distance larger than the window size from the current class making the *a posteriori* probabilities $P(\omega_j \mid x)$ zero for all j in the intervals: **0** $\leq$ **j** $<$ **(class-window)** and **(class+window)** $<$ **j** $\leq$ **24** with **(class-window)** $\geq$ **0** and **(class+window)** $\leq$ **24**.

All chromosomes are initially attached to the class with the highest *a posteriori* probability, still available, according to the Bayes' rule. The limit of two chromosomes per class is taken into account resulting in transfers where more than two chromosomes per class are found. This procedure continues until all chromosomes are either classified or rejected. A chromosome will be rejected if its *a posteriori* probability for the remaining classes is lower than a given minimum probability threshold (see Figure 9).

The second window, accessed through *Features for Classification* (see Figure 10), allows the user to adjust weights associated with each feature.

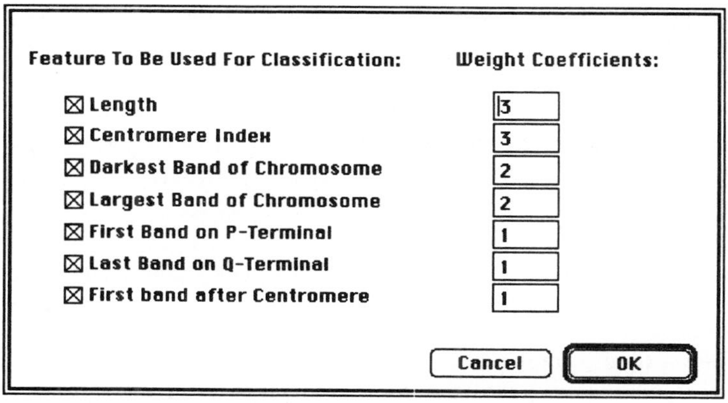

| Feature To Be Used For Classification: | Weight Coefficients: |
|---|---|
| ☒ Length | 3 |
| ☒ Centromere Index | 3 |
| ☒ Darkest Band of Chromosome | 2 |
| ☒ Largest Band of Chromosome | 2 |
| ☒ First Band on P-Terminal | 1 |
| ☒ Last Band on Q-Terminal | 1 |
| ☒ First band after Centromere | 1 |

Cancel     OK

**Figure 10**: The second window, *Features for Classification*, associated with the classification phase. The values entered by the user represent the weighting coefficients $\{\alpha_n\}$ as used in equation 5.

The default values used in this window were chosen for standard metaphase spreads as obtained through amniocentesis and Giemsa staining. Of particular importance in both of the windows associated with the classification phase is the ability to construct a database as well as a classifier that is "tuned" to the procedures and material within a given laboratory environment. That is, **Athena** can be used to construct a database *de novo*.

# 6. Results

In this section we present some of the results achieved with the techniques incorporated in **Athena**. Some of these results – such as the centromere detection accuracy and the classification accuracy – were generated in previous studies. The procedures, however, have been incorporated unchanged into this package and thus are representative of the results that can be achieved.

## 6.1. Centromere location

In this comparison we look at five techniques for determining the position of the centromere of banded chromosomes. The first technique is based upon the smallest distance across the chromosome [Visser, 1981]. The second technique uses the maximum deviation from the convex hull of the chromosome to the chromosome contour itself [Piper, 1981]. The third technique uses a polynomial fit to the width profile of the chromosome to find the centromere [Van Zee, 1974]. The profile is filtered and the deepest minimum used as the centromere. If no clear minimum exists – as for example in many acrocentric chromosomes – the inflection point is used to determine the centromeric position. The fourth technique, based upon the work of Lucas et al [1983], uses the difference between a "standard" width profile and the measured profile. The last method, the convex profile technique is based upon the deviation between the width profile and its convex hull [Piper et al, 1980].

In all of the comparisons, the chromosomes were first rotated and straightened (after Groen and van der Ploeg, 1979]. The details of the entire experiment are described in De Muinck Keizer [1984]. The results are shown in Table I.

In a prior study [de Muinck Keizer et al, 1983], an additional technique based upon the measurement of local curvature [Gallus, 1970] was examined. This technique led to an average accuracy of 15% and was thus discarded. The shortest distance method thus provides a significant improvement in the correct identification of the centromere position when compared to the other techniques. Its performance, however, remains below the 90% level. This indicates that, on the average, at least five chromosomes per metaphase would have to be interactively corrected. Whether this is acceptable may depend upon the nature of the user interface provided for the correction procedure.

**Table I**: Accuracy of the five centromere location schemes described in the text. 924 chromosomes [Van der Ploeg et al, 1974] were used in this study. A deviation of more than one pixel from the position indicated by the "expert" was considered as an error in centromere location.

| Chromosome Number | Shortest Distance | Convex Hull | Width Profile (1) | Width Profile (2) | Convex Profile |
|---|---|---|---|---|---|
| 1 | 76.5% | 52.9% | 73.5% | 64.7% | 85.3% |
| 2 | 80.0% | 71.4% | 77.1% | 62.9% | 91.4% |
| 3 | 92.3% | 43.6% | 74.4% | 59.0% | 78.9% |
| 4 | 100.0% | 72.1% | 83.7% | 69.8% | 90.5% |
| 5 | 95.6% | 73.3% | 95.6% | 44.4% | 91.1% |
| 6 | 89.2% | 75.5% | 83.8% | 86.5% | 91.9% |
| 7 | 87.8% | 82.9% | 90.2% | 87.8% | 87.8% |
| 8 | 92.3% | 97.4% | 97.4% | 76.9% | 97.4% |
| 9 | 81.1% | 78.4% | 94.6% | 81.1% | 89.2% |
| 10 | 89.7% | 92.3% | 82.1% | 74.4% | 87.2% |
| 11 | 78.6% | 73.2% | 85.7% | 61.9% | 76.2% |
| 12 | 83.3% | 69.0% | 83.3% | 73.8% | 83.3% |
| 13 | 95.3% | 92.1% | 41.9% | 83.7% | 66.7% |
| 14 | 97.5% | 72.5% | 17.5% | 85.0% | 52.6% |
| 15 | 97.3% | 92.1% | 34.2% | 92.1% | 72.7% |
| 16 | 83.3% | 88.4% | 86.0% | 76.7% | 100.0% |
| 17 | 82.1% | 71.8% | 35.9% | 84.6% | 82.1% |
| 18 | 92.5% | 92.7% | 36.6% | 92.7% | 95.1% |
| 19 | 53.5% | 82.9% | 48.8% | 67.4% | 83.3% |
| 20 | 69.2% | 86.5% | 74.4% | 69.2% | 94.9% |
| 21 | 86.8% | 39.4% | 34.2% | 71.1% | 50.0% |
| 22 | 62.5% | 63.9% | 57.5% | 35.0% | 61.8% |
| X | 96.6% | 82.2% | 100.0% | 89.7% | 96.6% |
| Y | 80.0% | 80.0% | 40.0% | 80.0% | 60.0% |
| **Average** | **85.1%** | **76.1%** | **67.9%** | **73.8%** | **81.9%** |

## 6.2. Classification accuracy

The complete procedure for classification described above has been tested on the same set of 924 chromosomes used for the centromere study. The test set was identical to the learning set because of the relatively small number of chromosomes per class ($\approx$20). The results are shown in Table II together with a comparison to Granum's WDD functions [Granum, 1981] on the same chromosomes.

**Table II**: Results of the classification of 924 chromosomes. In this experiment the test set equalled the learning set [ten Kate, 1983] for both classification techniques.

| Leiden Data Set | Laplace/Band Descriptors | WDD Functions |
|---|---|---|
| Error rate | 4.0% | 4.1% |
| Rejects | 0.0% | 1.6% |

Further, the two techniques – Laplace/band description versus WDD functions – were also compared on a much larger data set obtained from Lundsteen [1980]. The results are given in Table III.

Table III: Results of the classification experiment on the Lundsteen et al [1980] data set.

| Copenhagen Data Set | Laplace/Band Descriptors | WDD Functions |
|---|---|---|
| Error rate | 11.5% | 2.1% |
| Rejects | 0.0% | 0.1% |

In the *Laplace/Band Descriptors* technique, 7284 chromosomes were classified and the test set did *not* equal the learning set [ten Kate, 1983]. Bent chromosomes were classified but not straightened. In the *WDD* technique, 6985 chromosomes were classified and the test set *did* equal the learning set. Further, bent chromosomes were excluded from the classification procedure.

The use of the learning set as the test set in the testing of the WDD function approach and the exclusion of bent chromosomes mean that the value of 2.1% must be considered as highly optimistic. It is not possible for us to conclude that the technique we have implemented – a context sensitive classifier based upon bands identified by a form of Laplace filtering – is better than the WDD classifier. We can conclude, however, that it offers a reasonable accuracy. Further, the description developed by this classifier is much closer to the verbal description offered by cytogeneticists and embodied in the Paris convention [1971].

# 7. Conclusions

We have described in this article a software system for the (semi-)automatic analysis of metaphase spreads based upon a Macintosh II personal computer. This system takes full advantage of the hardware facilities in the computer: the 32 bit address space, the high speed disk, the floating point co-processor (required for chromosome rotation and straightening), the 8-bit deep display for color and grey levels, and the mouse-based user interface. As an indication of performance, the $512^2$ digitized metaphase image shown in Figure 11a and as a karyogram in Figure 11b requires a total of three minutes for processing. Less than fifty percent of this time represents "hands-off" automatic processing and the rest is used for the interactive correction of the segmentation and the final classification. In testing the newest version of Athena on a variety of metaphase spreads – including some produced in cell irradiation experiments – the automatic processing time was about 90 seconds and the time required for interaction ranged from an additional 90 seconds to an additional 270 seconds.

Athena

Date : Monday, April 4, 1988; 12:42:15
Image Name/ID : chromo1.1
Comments : The metaphase spread "chromo1.1" printed after contrast stretching.

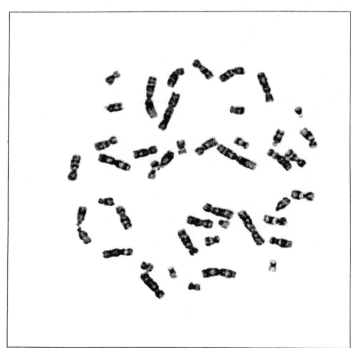

Figure 11a

Athena

User : Lucas van Vliet
Date : Monday, April 4, 1988; 16:15:11
Slide ID : Unknown
Metaphase Location : X-Coordinate = 0.0        : Y-Coordinate = 0.0
Comments : Female karyotype without any visible abnormalities. The original metaphase image is stored on disk in the image file named "chromo1.1".

Figure 11b

Because of the flexibility of this system – the ability to build a database with "fine-tuning" of the classification, analysis and classification parameters – we feel that it should be possible to extend it easily to a variety of cytogenetic material such as corionic villi, bone marrow, and solid tumors. We look forward to reporting in the future about the developments in **Athena**.

*Acknowledgement.* This work was partially supported by the Commission of the European Communities through the Medical and Health Research Program, project number II.1.1/13. We would also like to acknowledge the help and advice of our colleagues in Leiden, Copenhagen, Edinburgh, and Livermore.

# References

1.    APDA – Apple Programmer's Development Association (1988), *Macintosh Programmer's Workshop Reference – Version 2.0*, Renton, Washington.

2.    Apple, *Inside Macintosh Vol. I-III*, Addison-Wesley Publishing Company Inc., Reading, Massachusetts, 1985.

3.    Apple, *Inside Macintosh Vol. IV*, Addison-Wesley Publishing Company Inc., Reading, Massachusetts, 1986.

4.    Apple, *Inside Macintosh Vol. V*, Addison-Wesley Publishing Company Inc., Reading, Massachusetts, 1987.

5.    Bishop RP and Young IT (1977), *The automated classification of mitotic phase for human chromosome spreads*, Journal of Histochemistry and Cytochemistry, Vol. 25:7, pp. 730-740.

6.    Caspersson T, Zech L, Johansson C (1970a), Differential banding of alkylating fluorochromes in human chromosomes, Experimental Cell Research, Vol. 60, pp. 315-319.

7.    Caspersson T, Zech L, Johansson C and Modest EJ (1970b), Identification of human chromosomes by DNA-binding fluorescent agents, Chromosoma, Vol. 30, p. 215.

8.    de Muinck Keizer M, Groen FCA, Young IT, Smeulders AWM (1983), *An objective comparison of various centromere determination techniques*, Proceedings of the Vth European Chromosome Analysis Workshop, Heidelberg, West Germany.

9.    de Muinck Keizer M (1984), *An objective comparison of various centromere determination techniques*, Master's thesis, Department of Applied Physics, Delft University of Technology, (in Dutch), (Thesis supervisors: F.C.A. Groen and I.T. Young).

10.   Deriugen NG (1957), *The power spectrum and correlation function of the television signal*, Telecommunications, Vol. 7, pp. 1-12.

11.   Duda RO and Hart PE, *Pattern Classification and Scene Analysis*, John Wiley – Interscience, 1973.

12.   Duin RPW (1983), *Interactive image processing in a multi-user environment*, in: Computing Structures for Image Processing, Ed. MJB Duff, Academic Press, London, Chapter 8.

13.   Gallus G and Neurath PW (1970), *Improved computer chromosome analysis incorporating preprocessing and boundary analysis*, Physics in Medicine and Biology, Vol. 15, pp. 435-445.

14.   Granum E, Gerdes T, Lundsteen C (1981), *Simple weighted density distributions, WDD's for discrimination between G-banded chromosomes*, Proceedings of the IVth European Chromosome Analysis Workshop, Edinburgh.

15.  Granlund GH, Zack GW, Young IT, Eden M (1976), *A technique for multiple-cell chromosome karyotyping*, Journal of Histochemistry and Cytochemistry, Vol. 24:1, pp. 160-167.

16.  Groen FCA, Verbeek PW, Zee GA van der, Oosterlinck A (1976), *Some aspects concerning computation of chromosome banding profiles*, in Proceedings of the IIIrd International Joint Conference on Pattern Recognition, Coronado, California, pp. 547-550.

17.  Groen FCA and van der Ploeg M (1979), *DNA cytophotometry of human chromosomes*, Journal of Histochemistry and Cytochemistry, Vol. 27, pp. 435-440.

18.  Groen FCA, Young IT, Ligthart G (1985), *A comparison of different focus functions for use in autofocus algorithms*, Cytometry, Vol. 6, pp. 81-91.

19.  Krusemark S and Haralick RM (1983), *An operating system interface for transportable image processing software*, Computer Vision, Graphics, and Image Processing, Vol. 23, pp. 42-66.

20.  Ledley RS (1964), *High speed automatic analysis of biomedical pictures*, Science, Vol. 146, pp. 216-223.

21.  Ledley RS, Rotolo LS, Golab TJ, Jacobsen JD, Ginsberg MD, Wilson JB (1965), *FIDAC – Film input to digital automatic computer and associated syntax directed pattern recognition programming system*, in Optical and Electro-Optical Information Processing, Ed's. Tippet JI, Beckowitz D, Clapp L, Koester C, Vanderburgh A, MIT Press, Cambridge, Mass., pp. 591-631.

22.  Lucas JN, Gray JW, Peters DC, Van Dilla MA (1983), *Centromeric index measurement by slit-scan flow cytometry*, Cytometry, Vol. 4:2, pp. 109-116.

23.  Lundsteen C, Philip J, Granum E (1980), *Quantitative analysis of 6985 digitized trypsin G-banded human metaphase chromosomes*, Clinical Genetics, Vol. 18, pp. 335-370.

24.  Mayall BH, Carrano AV, Rowley JD (1974), *DNA cytophotometry of chromosomes in a case of chronic myelogeneous leukemia*, Cancer Research, Vol. 37, pp. 3590-3593.

25.  Mayall BH, Carrano AV, Golbus MS, Conte FA, Epstein CJ (1977a), *DNA cytophotometry in prenatal cytogenetic diagnosis*, Clinical Genetics, Vol. 1, pp. 273-276.

26.  Mayall BH, Carrano AV, Moore DH, Rowley JD (1977b), *Quantification by DNA-based cytophotometry of the 9q+/22q-chromosomal translocation associated with chronic myelogeneous leukemia*, Cancer Research, Vol. 37, pp. 3590-3593.

27.  Mayall BH, Carrano AV, Moore DH, Ashworth LK, Bennet DE, Mendelsohn ML (1984), *The DNA-based human karyotype*, Cytometry, Vol. 5, pp. 376-385.

28.  Mendelsohn ML, Hungerford DA, Mayall BH, Perry BH, Conway TJ, Prewitt JMS (1969), *Computer-oriented analysis of human chromosomes. II. Integrated optical density as a single parameter for karyotype analysis*, Annals of the New York Academy of Science, Vol. 157, pp. 376-393.

29.  Mendelsohn ML and Mayall BH (1972), *Computer-oriented analysis of human chromosomes. III. Focus*, Computers in Biology and Medicine, Vol. 2, pp. 137-150.

30.  Mendelsohn ML, Mayall BH, Bogart E, Moore DH, Perry BH (1973), *DNA content and DNA-based centromeric index of the 24 human chromosomes*, Science, Vol. 179, pp. 1126-1129.

31.  Mendelsohn ML, Bennet DE, Bogart E, Mayall BH (1974), *Computer-oriented analysis of human chromosomes. IV. Deoxyribonucleic acid based centromeric index*, Journal of Histochemistry and Cytochemistry, Vol. 22, pp. 554-560.

32.  Neurath PW et al (1965), *Human chromosome analysis by computer – an optical pattern recognition problem*, Annals of the New York Academy of Science, Vol. 128, pp. 1013-1028.

33.  Paris Conference (1971), *Standardization in human cytogenetics*, Birth Defects: Original Article Series, National Foundation – March of Dimes, New York, Vol. VIII:7.

66      L.J. van Vliet, I.T. Young, T.K. ten Kate, B.H. Mayall, F.C.A. Groen, R. Roos

34.    Piper J, Granum E, Rutovitz D, Ruttledge H (1980), *Automation of chromosome processing*, Signal Processing, Vol. 2, pp. 203-221.

35.    Piper J (1981), *Finding chromosome centromeres using boundary and density information*, in Digital Image Processing, Ed's. J-C Simon and RM Haralick, Dordrecht, The Netherlands, pp. 511-518.

36.    Ploeg M van der, Duyn P van, Ploem JS (1974), *High resolution scanning densitometry of photographic negatives of human metaphase chromosomes, Parts I and II*, Journal of Histochemistry and Cytochemistry, Vol. 42, pp. 9-46.

37.    Preston K (1983), *Progress in image processing languages*, in: Computing Structures for Image Processing, Ed. MJB Duff, Academic Press, London, Chapter 13.

38.    Ridler TW and Calvard S (1978), *Picture thresholding using an interactive selection method*, IEEE Transactions on Systems, Man and Cybernetics, Vol. SMC-8:8, pp. 630-632.

39.    Schreiber WF (1956), *Third-order probability distribution of television signals*, IRE Transactions on Information Theory, Vol. IT-2:3, pp. 94.

40.    Tamura H, Sakane S, Tomita F, Yokoya N, Kaneko M, and Sakaue K (1983), *Design and implementation of SPIDER – A transportable image processing software package*, Computer Vision, Graphics, and Image Processing, Vol. 23, pp. 273-294.

41.    Ten Kate TK, Groen FCA, Young IT, van der Ploeg M, Pearson PL, Gerdes T, Lundsteen C (1983), *The Delft classification technique on some difficult chromosome classes*, Proceedings of the Vth European Chromosome Analysis Workshop, Heidelberg, West Germany.

42.    Ten Kate TK (1985), *Design and implementation of an interactive karyotyping program in C on a Vicom digital image processor*,  Master's thesis, Department of Applied Physics, Delft University of Technology (Thesis supervisors: F.C.A. Groen and I.T. Young).

43.    Tjio JH and Levan H (1956), *The chromosome number in man*, Hereditas, Vol. 42, pp. 1-6.

44.    Visser RT (1981), *Classification of banded chromosomes using a priori cytologic knowledge*, Master's thesis, Department of Applied Physics, Delft University of Technology, (in Dutch), Master's thesis, Department of Applied Physics, Delft University of Technology (Thesis supervisor: F.C.A. Groen).

45.    Vliet LJ van and Verwer BJH (1988), *A contour processing method for fast binary neighborhood operations*, Pattern Recognition Letters, Vol. 7, pp. 27-36.

46.    Vliet LJ van, Young IT, Beckers ALD (1988), *A non-linear Laplace operator as edge detector in noisy images*, accepted by Computer Vision, Graphics, and Image Processing.

47.    Vossepoel AM, Smeulders AWM, Van de Broek K (1979), *DIODA: Delineation and feature extraction of microscopical objects*, Computer Programs in Biomedicine, Vol. 10, pp. 231-244.

48.    Young IT, Groen FCA, Dorst L, Geerlings AC, Jovin T, Arndt-Jovin DJ (1983), *Comparative imaging and quantification of DNA species in chromosomes*, Proceedings of the Vth European Chromosome Analysis Workshop, Heidelberg, West Germany.

49.    Young IT and Roos R (1988), *Acuity: Image analysis for the personal computer*, in: Pattern Recognition in Practice, Vol. III, Ed's. ES Gelsema and LN Kanal, North Holland Press, Amsterdam.

50.    Zack GW, Spriet JA, Latt SA, Granlund GH, Young IT (1976), *Automatic detection and localization of sister chromatid exchanges*, Journal of Histochemistry and Cytochemistry, Vol. 24:1, pp. 168-177.

51.    Zee GA van (1974), *Automated chromosome analysis of Feulgen stained specimens*, Master's thesis, Department of Applied Physics, Delft University of Technology, (in Dutch), Master's thesis, Department of Applied Physics, Delft University of Technology (Thesis supervisor: F.C.A. Groen).

# Recent Advances in Chromosome Analysis with the Chromoscan

P. Malet, A. Geneix, P. Bonton, L. Grouche, P. Huygue

**Summary**

The development of metaphase-finding and karyotyping machines presently constitutes one of the principal areas of advancement of human cytogenetic techniques. The rapidly growing demand for karyotypes, especially for pre-natal diagnosis, has led current research to aim at reducing the time and cost of these tests as well as increasing their precision. We present here the latest improvements of the Chromoscan, a karyotyping machine that we have developed and adapted to hospital use [1-8].

## 1. Description

The Chromoscan (fig 1, 2) is made up of the following components:

1. IBM PC/AT computer equipped with:
   - high resolution 1024×1024 digitizer with 256 levels of grey intensity
   - serial communications port
   - parallel communications port
   - monochromome screen
   - 20 or 30 Mbyte hard disk
   - math co-processor
2. Thomson TAV1030 video camera
3. High resolution video screen
4. Microsoft mouse
5. Honeywell VGR 5000, or Laser-technics printer
6. 20 Mbyte streamer tape back up system with software, or optical disk.
7. I3M network
8. Software
   - DOS
   - Chrodiag karyotyping software
   - Chrodiag II karyotyping software
   - Chromodiag expert system software

**Automation of Cytogenetics**   Editors: C. Lundsteen J. Piper
© Springer-Verlag Berlin Heidelberg New York 1989

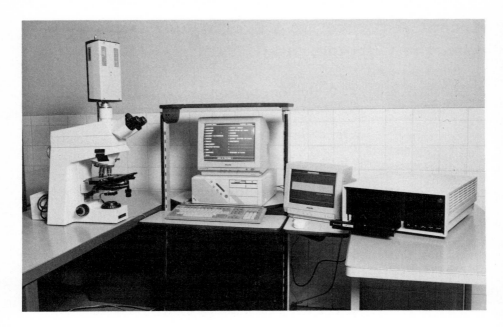

**Fig. 1.** The Chromoscan

## 2. Interactive karyotyping

### 2.1. Pretreatment and enhancement of the image

The quality of the image transmitted by the microscope varies from one mitosis to another and the levels of grey intensity corresponding to the bands are consequently more or less easy to identify. We have therefore developed a stage of pretreatment in order to improve the visualization of chromosomal heterostructures.

The information received by the system is encoded using a scale of 256 levels of grey intensity. But the information concerning the chromosomes is located in a small part of this scale toward the low levels. The pretreatment consists of spreading out the scale of grey intensities found in the chromosomes over the entire scale of the image processing system. The mode corresponding to the background of the preparation and the chromosome scale are automatically determined from the histogram of the image received. Spreading out this small range of pertinent levels stretches contrasts; the background of the preparation becomes more homogeneous, less present, and the banding patterns are more visible. A greater number of metaphase spreads are therefore exploitable. This enhancement can be performed on an entire image or on

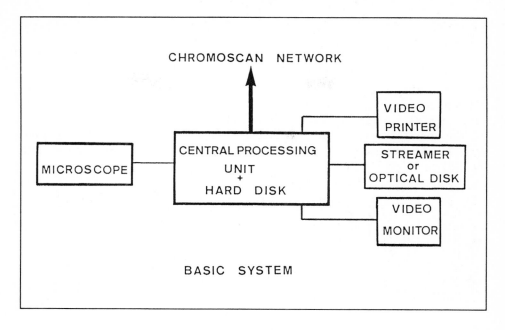

**Fig. 2.** Chromoscan architecture

individual chromosomes; the band sharpening facilitates in certain cases the chromosome identification.

## 2.2. Automated determination of contours

The enhanced version of the image is presently used to carry out a digitalization, a propagation of a fictively determined level of grey intensity to identify the background, the filling in of "holes", and the elimination of particles whose morphological parameters do not correspond to those of the desired structures. The digital image thus obtained instantaneously provides all the contours of the chromosomes that can then be isolated.

This method allows the user not only to obtain the contours immediately, but also to get them with a greater precision than with the interactive method. In the case of touching and overlapping chromosomes, however, the operator still needs to intervene with the mouse.

## 2.3. Isolating the chromosomes

To isolate the chromosomes and count them, the method is one classically used in the field of mathematical morphology. The image is scanned vertically and horizontally and a level of grey intensity is assigned to each chromosome particle as a label. Then, using this image with 46 levels of grey intensity, the operator needs only choose (by thresholding) the level of grey intensity necessary to isolate a given chromosome and superimpose this image on the original image to determine the real chromosome.

## 2.4. Counting and classifying

First the chromosomes are counted by the operator. The mouse then allows the axis of each chromosome to be located by choosing two points, one on the short arm, the other on the long one. The chromosome is then transferred by the operator to the classifying board which can be visualized upon request (fig. 3). As was indicated above, touching or overlapping chromosomes must be outlined with the mouse.

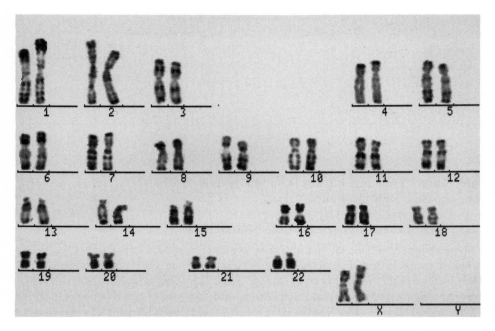

**Fig. 3.** A karyotype displayed on the high resolution monitor

## 2.5. Other functions

The following functions are available:
- Straightening chromosomes
- Colour assignment to chromosome bands and to background
- Electronic zoom/enlargement
- Vertical translation of chromosomes in order to align centromeres.
- Inversion of chromosomes
- Erasing artefacts or chromosomal fragments

– Simultaneous display of multiple metaphases on the screen.
– Inscription of administrative and medical data.

The number of chromosomes on the screen and the distance between chromosome groups can be easily programmed. This function allows karyotyping of animal and plant species.

Tables of chromosomes of the same class group belonging to different metaphases can also be created.

The user has the option of producing a negative image with a pseudofluorescent aspect.

## 2.6. Printing the karyotype

The karyotype is printed on a high definition printer.

## 2.7. Storage

The unclassified metaphases and the karyotypes are stored in the archives on a streamer or on an optical disk.

# 3. Automated karyotyping

Two methods can be envisaged:

1. A statistical study of a large number of chromosomes coming from a cytogenetic library can be performed. This is slow and the data obtained depends on the techniques used which are variable from one laboratory to another.
2. Shape recognition can be approached using a model. The principal criteria of identification are the following: length of the chromosome, position of the centromere, number and position of the bands, diverse landmarks (secondary constriction, etc.).

We have opted for this second technique.

The automated classification of the chromosomes is preceeded by a phase of pretreatment which allows a number of metaphases to be studied (fig 4a-f).

The identification phase begins with the determination of the length of chromosomes and the indirect location of the centromeres using a line by line scanning that allows the zones of separation and fusion of chromosomal arms to be defined. These latter are distinguishable by their level of grey intensity and the space that separates them. The subsequent determination of the centromeric index will be the first element of the chromosome identification.

The chromosomes are then "straightened out". A preclassification is carried out according to their size and centromeric index.

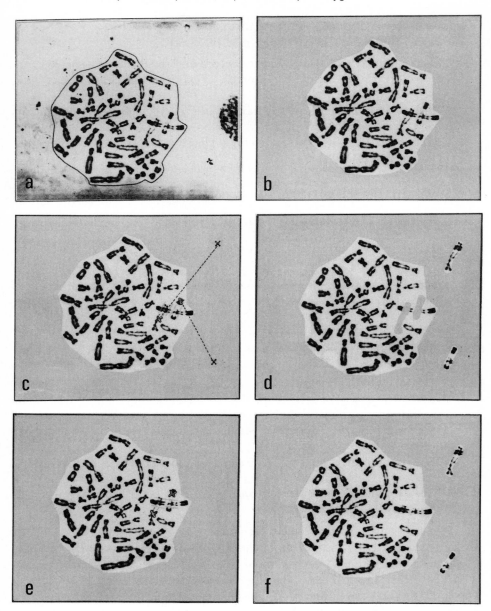

**Fig. 4.** Prelimary stages to automated karyotyping. (a) Delineation of the area to be studied. (b) Elimination of the background. (c) Designation of the chromosomes to be isolated. (d) Transfer of the chromosomes out of the metaphase. (e) Erasure of duplicated chromosomes. (f) Result.

## Utilization of densitometric curves for chromosome identification

A more precise classification is finally made after examining the banding patterns and drawing up densitometric curves. A chromosome densitometric curve is characterized by "peaks" and "valleys" of varying width. We have chosen the following procedure for the Chromoscan. A narrow window on the slide of 3 pixels by 3 pixels ($0.420\mu m \times 0.305\mu m$) orientated perpendicularly to the chromosome's axis, scans the image line by line from one telomere to another. The value obtained for the densitometric curve corresponds to the sum of all the points studied on the line. The sum of all the chromosomal pixels is calculated, each one weighted according to the intensity of its greyness calculated as shown (fig 5). This value is memorized; its variations on the length of the chromosome will define the densitometric curve (fig 6).

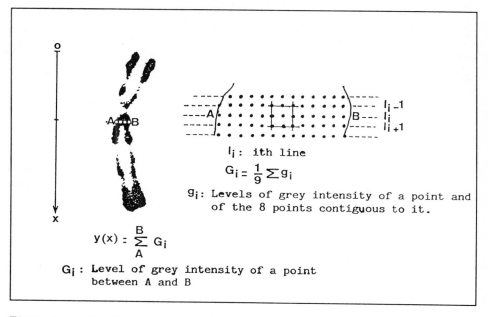

**Fig. 5.** Computing the density profile. Left, the profile is the sum of "grey values" at each point. Right, the definition of the "grey value" at a point.

We have also provided for the possibility of simultaneously visualizing 4 densitometric curves next to the corresponding chromosomes. This choice facilitates the comparison of curves, especially in the case of reciprocal translocation. The operator uses the mouse to designate the chromosome to be analyzed on the karyogram. The densitometric curve is then displayed in a frame in the center of the screen, next to the corresponding chromosome. The karyotype appears upon request and a second chromosome can be designated for analy-

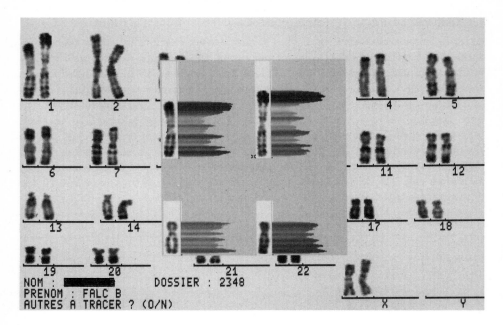

**Fig. 6.** Shaded density profiles displayed on the high resolution monitor.

sis. It is displayed next to the first one. A total of 4 chromosomes can thus be displayed at a time (fig 6). If a fifth chromosome is designated, it will take the place of the first one and so on.

Densitometric curves for individual chromosome areas can also be elaborated.

The surface situated under the curves incidentally presents the same variations in intensity of greyness as the chromosome. This amplifies the differences for the eye without, of course, providing supplementary information.

The position of the centromere is displayed as a sharp drop in the curve.

One problem that will obviously arise is that of overlapping or touching chromosomes. To resolve this, an interactive mode proves to be necessary. Rearranged chromosomes can be detected by their atypical densitometric profiles. Formerly, this was done by ocular observation of the curves, without automated analysis.

The operator conserves the possibility of correcting the final classification.

Moreover, a statistical study based on a large number of mitoses can allow us to identify the stable factors of the reverse banding pattern and to elaborate a "key-lock" system for chromosome identification.

# 4. The Chromoscan network

The use of a local network for chromosomal analysis complies with two objectives:

- The first is the shared use of the printer, the mass storage, the modem, etc., in order to reduce the cost of analyses. The price of the network is largely compensated by the material economized.
- The second is the elaboration of a coherent system that allows the flow of information among Chromoscan stations and with the exterior.

The criteria for choosing a network are the following:

- High general performance (high speed transmission (10 Mbaud) and rapid processing).
- Capacity to be updated.
- Facility of connection with other networks.

In compliance with the above criteria we have opted for the 3 COM network. It consists of an electronic card and software. A coaxial cable allows it to be connected to different stations.

Three programs are used in a manner totally transparent to the user:

- A 3N service that consists of a data base relative to the network users.
- A 3F service that defines the modalities of the memory allocation among the users and establishes the connections between the computers. It also disposes of approximately 20 process controls. The connection with the mass storage is automatic.
- A 3P service that controls the allocation of the use of printer, optical disk, etc., and establishes the logical connections between the computers and the material. It also manages a waiting list and defines the priority of print-outs.

The Chromoscan network expands the basic Chromoscan station to multiple users to simultaneously and independently create karyotypes. All users share the same image data base and a high quality image printer. Additional stations may be added at lower cost.

# 5. Multiple scanning system

The system (fig 7) consists of a Chromoscan karyotyping workstation and two or more metaphase scanning stations where metaphases are searched by an operator.

Each metaphase scanning station consists of a microscope, a camera and its appropriate microscope adaptor, a keyboard/lap-top computer and a monochrome monitor. The purpose of the monitor in the scanning station is to allow the operator to see and focus the metaphases found manually.

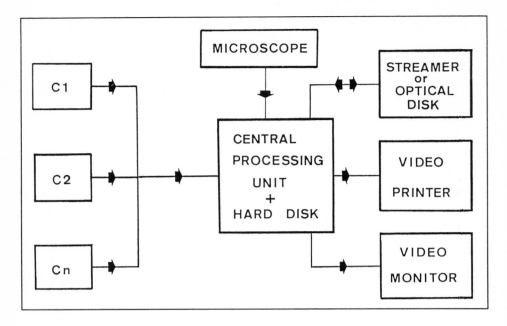

**Fig. 7.** Multiple camera system

Once a metaphase has been found, the operator may send it to the karyotyping station and store it in the hard disk of the karyotyping computer.

Images can only be enhanced with the karyotyping workstation.

The keyboard/lap-top computer is only used to enter general information about the case being worked on (patient identification, report, type of tissues, date of analysis, initials of the operator at the scanning station).

The multiscanning system enables several microscopes to be connected to a single karyotyping workstation. It provides several cytogeneticists with direct access to the karyotyping station. While one cytogeneticist performs a karyotype others can search for metaphases at their own microscope and input their images to the karyotyping workstation.

## 6. Signal evaluation after in situ hybridization

In order to map the genes, *in situ* hybridization is usually followed by a statistical study of the distribution of the silver grains. This method is long and tedious. We cannot distinguish one high signal from another with the statistical test of frequencies. Each grain has the same value, even in cases of multicopy genes.

We have drawn up a procedure for the quantification of the levels of grey intensity using the Chromoscan. The intensity of the radioactive signal is

**Table 1a,b.** *In situ* hybridisation results.

| Number of the mitosis | Number of the band | Intensity of the band |
|---|---|---|
| 1 | 03P00 | 5608 |
| 1 | 12P00 | 1680 |
| 1 | 03Q28 | 716 |
| 1 | 06P00 | 476 |
| 2 | 03P00 | 8827 |
| 2 | 09P00 | 3565 |
| 3 | 03P00 | 3064 |
| 3 | 12Q24 | 1152 |
| 3 | 17P11 | 742 |
| 4 | 03P00 | 5381 |
| 4 | 16P11 | 1268 |
| 4 | 07P00 | 933 |
| 4 | 05P14 | 510 |
| 5 | 02P00 | 5308 |
| 5 | 03Q26 | 1264 |
| 5 | 03P00 | 1106 |
| 5 | 0XQ21 | 250 |
| 6 | 03P00 | 13152 |
| 6 | 05P14 | 1233 |
| 6 | 04P11 | 494 |
| 7 | 03P00 | 9071 |
| 7 | 11P15 | 1783 |
| 7 | 09P23 | 762 |
| 7 | 11P11 | 506 |

| Number of the band | Average of the intensities | Variance |
|---|---|---|
| 03P00 | 6587 | 1.4845 |
| 12P00 | 240 | 0.0308 |
| 03Q28 | 102 | 0.0056 |
| 06P00 | 68 | 0.0024 |
| 09P00 | 509 | 0.1390 |
| 12Q24 | 165 | 0.0145 |
| 17P11 | 106 | 0.0060 |
| 16P11 | 181 | 0.0176 |
| 07P00 | 133 | 0.0095 |
| 05P14 | 249 | 0.0159 |
| 02P00 | 758 | 0.3082 |
| 03Q26 | 181 | 0.0174 |
| 0XQ21 | 36 | 0.0006 |
| 04P11 | 71 | 0.0026 |
| 11P15 | 255 | 0.0348 |
| 09P23 | 109 | 0.0063 |
| 11P11 | 72 | 0.0028 |

evaluated according to the sum of the values assigned to all the grey intensity levels in each marked area. Each chromosome band in every mitosis is thus measured. For each band the average value and the statistical variation appear numerically. The chi-square test for the frequencies of the signal and Student's $t$-test for the averages of the intensities are applied to reveal the most probable locus of the gene.

For example, we have tested a multicopy gene and the locus appeared clearly near the #3 centromere. In this case, Student's $t$ (done with few data) had shown a significant difference ($\alpha = 1\%$) between 3P00 and the band that was nearest in intensity at 2P00. The chi-square test on the frequencies shows an even more significant difference ($\alpha < 0.1\%$), and this for every band compared with 3P00 (table 1).

This variation is due, first of all, to the fact that the chi-square test is very effective. Secondly, the grain count alone does not take the size of the grains into account, and this is reflected in the results.

In cases like the one above, where both tests are significant, the results agree and we can conclude without difficulty: the locus is probably at 3P00.

In certain cases the conclusion is less obvious: Student's $t$-test is significant but the chi-square test is not, or *vice versa*. In the first case, this indicates that one or several loci might be more frequent than the locus that is globally the most marked. The judgement must be made cautiously, in particular by taking into account the nature of the probes: repetitive or single copy.

If the chi-square test is significant and Student's $t$-test is not, this indicates that the most frequent locus is not globally the most marked. In a case like this, it seems that only an intense background noise can explain the result. The judgement must be made with caution.

It is advisable to do both tests. Since the system is automated, these two operations are carried out without difficulty.

# 7. Results and conclusion

The Chromoscan has been used routinely for 2 years in a hospital laboratory. 1200 prenatal tests were carried out during this period (amniotic fluid, chorionic villi, foetal blood). The chromosomes of 16 mitoses were compared for each patient. In each case, 4 mitoses were classified and visualized in the form of hardcopies. The average time for prenatal diagnosis required 60 to 80 minutes. For all the cases examined taken together, the cost was reduced by a factor of five. The precision is comparable to that of conventional methods.

Our interactive karyotyping system presents the following advantages: reduction of cost and analysis time; all processing facilities related to image digitalization; easy integration of additional functions; interaction with the operator possible at any time; and access to various fields of quantitative cytogenetics.

Higher performance, less expensive video cameras, computers, image digitizers and storing devices will allow complementary advances in the near future.

The memory capacity of computers is presently being miniaturized and augmented. Nevertheless, the presence of a screen and a microscope will continue to impose a minimum amount of required space that cannot be reduced.

The use of digitizers and printers with a high number of grey levels is necessary. However, the limit of resolution will obviously remain dependent on the quality of the preparation examined. This latter should be specified in the case of evaluation tests.

Specific programs may be elaborated in several domains such as the detection of chromosome aberrations, heterochromatin quantification, the study of prometaphasic chromosomes, or sister chromatid exchanges. The Chromoscan can also easily be adapted for use in animal and plant cytogenetics.

*Acknowledgements.* We would like to express our gratitude to B. Perissel, G. Rongier, B. Le Corvaisier, and Y. Faugeras for their collaboration, and to Chromoscan France.

# References

1     Malet P, Geneix A, Bonton P, Catinot PH, Perissel B. Un nouveau système d'analyse chromosomique par traitement d'images: Le Chromoscan. C.R. Soc. Biol. 181:216-220 (1987)
2     Catinot PH, Malet P, Bonton P, Geneix A, Perissel B. Un système vidéo interactif pour l'analyse du caryotype humain. La Presse Médicale 16(2)80-81 (1987)
3     Malet P, Geneix A, Perissel B, Bonton P, Grouche L. Le caryotype outil genetique. Aspects récents de l'analyse chromosomique. Biomequip 13:25-33 (1987)
4     Malet P, Geneix A, Bonton P, Grouche L, Perissel B. Apports récents de l'analyse par traitement d'images aux études chromosomiques. Bull. Soc. Fr. Genet. 2:18 (1987)
5     Malet P, Bonton P, Geneix A. L'analyse chromosomique par traitement d'image: un nouvel atout dans la prévention et le diagnostic des maladies d'origine génétique. Auvergne Sciences Bull. de l'Adasta 5:3-7 (1988)
6     Malet P, Geneix A, Bonton P, Perissel B, Turchini MF, Charbonne F, Jaffray JY, Grouche L, Turchini JP. Les nouvelles technologies cytogénétiques et leurs applications médicales. Semaine des Hopitaux 23:1576-1586 (1988)
7     Geneix A, Malet P, Bonton P, Perissel B. Un système interactif d'analyse chromosomique. Cytologia (Japan) 53:509-515 (1988)
8     Geneix A, Malet P, Bonton P, Grouche L, Perissel B. Image Processing in Human Cytogenetics: New Steps Toward Quantification. Karyogram (USA), 14:45-49 (1988)

# Proposal for Evaluation of Cytogenetic Laboratories in Order to Plan the Installation and Determine the Cost Effectiveness of Automated Chromosome Analysis Systems

J.M. García-Sagredo

## Summary

This paper proposes a method for objective evaluation on a national basis of cytogenetic laboratories' need for automated chromosome analysis systems, including evaluation of the most appropriate type of system, based on an analysis of the size, staffing structure, facilities and throughput of each laboratory. The evaluation method has been developed from experience of seven laboratories in Spain, and the results of applying the method to 14 laboratories of various different sizes and structure are presented.

## 1. Introduction

From 1957 with the introduction of cytogenetics techniques into routine medical practice, the utilization of chromosome analysis has increased rapidly, principally in the last 10 years; but these techniques are expensive, and time and personnel consuming. For these reasons, several attempts at automation of some technical procedures in chromosome analysis have been made in the last decades, although it is only in recent years that automatic systems (AS) for chromosome analysis have been commercially available and had a real effectiveness, becoming useful in cytogenetic laboratories.

Due to the novelty and price of AS for cytogenetic studies, it is difficult to convince hospital administrators about their advantages and cost effectiveness. These difficulties arise for the following reasons:

a) poor knowledge of genetics and its procedures
b) efficiency of genetics procedures in the hospital not always known
c) small number of studies in genetic labs when compared to hematologic or biochemical ones
d) the cost efficiency of the AS is not clearly elucidated.

Sometime, especially in countries with a national health system, it is easier to acquire such systems if a plan for installation is made globally than it is to get them for an individual lab by a personal or individual decision. In any case, prior planning based on present and future needs, and knowledge of the cost

**Automation of Cytogenetics**   Editors: C. Lundsteen J. Piper
© Springer-Verlag Berlin Heidelberg New York 1989

efficiency of automation in some chromosome analysis procedures, allows either individual acquisition (without losing the general point of view of the national health system), or a planned acquisition of several AS to cover the established demands.

To obtain a good cost effectiveness when the introduction of AS is planned, it is not only necessary to evaluate such systems carefully (if possible by a widely accepted test), but it is also necessary to evaluate cytogenetic labs, taking into account two general points:

1) the number of samples processed
2) the type of samples

But cytogenetic labs do not always have the same structure, and there are differences from one country to another. In the case of Spain, the characteristics and staff structure of the laboratories made necessary an evaluation which considers the following points in order to obtain better results:

a) few and poorly qualified technical people
b) most of the technical work is carried out by medical people who share it with their clinical work
c) small staff and an increasing demand
d) technical equipment sometimes obsolete.

With these facts, in the case of Spain (and other countries with similar characteristics) it is necessary to consider the following points when evaluating cytogenetic labs:

1) number of cases (weighted according to their difficulty).
2) type of samples: peripheral blood (PB), amniotic fluid (AF), chorium villi (CVs), bone marrow (BM), etc.
3) number of technicians.
4) ratio of doctors (with technical duties) and technical people
5) lab material and equipment
6) type of hospital (regional, local, etc.)

## 2. Methods

The aim of this proposal is to obtain an objective evaluation of the labs which will indicate high or low cost effectiveness of the AS through a high or low score, together with a qualitative evaluation directed to different types of AS. Before proceeding to describe the proposed methodology, some preliminary points should be considered:

a) The formulas proposed and the figures chosen as indicators between different characteristics (especially in the qualitative evaluation of cytogenetic labs) have been developed and evaluated using a set of seven Spanish cytogenetic laboratories chosen because of their different structure, and which cover the spectrum of possibilities of lab structure which can be found in Spain.

b) When a cytogenetic lab is evaluated, the methodology is directed to establishing the efficiency of an AS in it; nevertheless all the procedures in the lab were evaluated, although it might not be possible to automate them all (e.g. culture harvest, tripsinization, etc.). This is because these procedures are time and personnel consuming and it is not possible to take them out of the general lab context. Thus people who are "liberated" by a procedure becoming automated can be employed in other procedures of the lab; or, on the contrary, if an automated procedure makes the chromosome analysis faster, obviously this can influence almost all the procedures in the lab.

For cytogenetic laboratories, we consider two types of evaluation. One is quantitative, by which we obtain a score of AS cost effectiveness within a certain laboratory. The other is a qualitative evaluation which indicates the type and performance requirement of AS for a particular lab.

## 2.1. Quantitative evaluation

For quantitative evaluation we consider six points (Figure 1). These items are evaluated and scored separately according to different methodologies:

**Number of samples:** It is well known, for comparison or evaluation purposes, that there are differences in the time taken and difficulty of different samples for chromosome analysis. Because of this, the number of samples per year analysed by the lab is corrected by giving a weight according to the technical difficulty of the type of sample. The weights are used according to the method reported by Lubs et al (1986) and are the following (peripheral blood 1; amniotic fluid 1.5; chorium villi 1.3; bone marrow 2; peripheral blood from leukemia 1.5; and fibroblasts 1.5).

The weights make reference to the technical difficulty of the whole procedure for chromosome analysis from taking the sample in the beginning until finishing the karyogram report. The reason for this, as explained above, arises because of the difficulty of isolating the individual procedures, which is why they are seen as a complete process involving all the staff members of the lab (see the following points).

The score of this item goes from 0 (less than 500 samples/year) to a maximun of 5 (more than 1,500 cases/year).

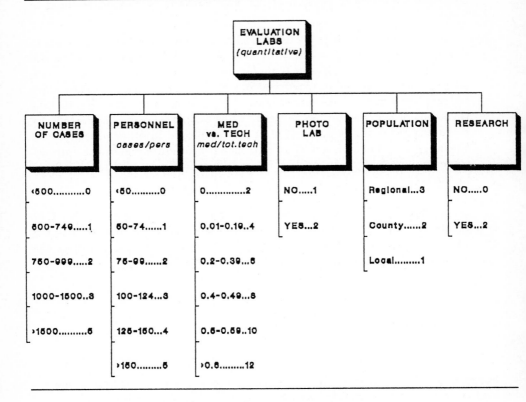

**Fig. 1.** Quantitative laboratory evaluation procedure

| Samples/year | Score |
|---|---|
| <500 | 0 |
| 500 – 749 | 1 |
| 750 – 999 | 2 |
| 1000 – 1500 | 3 |
| >1500 | 5 |

**Staff:** We consider the ratio of the number of samples to the number of staff members as an indicator of efficiency of personnel. The corresponding score varies from 0 to 5.

| Samples/person/year | Score |
|---|---|
| <50 | 0 |
| 50 – 74 | 1 |
| 75 – 99 | 2 |
| 100 – 124 | 3 |
| 125 – 150 | 4 |
| >150 | 5 |

**Proportion between highly qualified people versus technicians:** We consider that, as in the case of Spain, if the number of technicians is low (which means that many highly qualified people are using their time on technical analysis) the cost effectiveness of an AS will be higher because it substitutes a more expensive and specialized time.

For us, this point is important in the evaluation of the performance of a laboratory, and the score range varies from 2 to 12. The score is computed by a formula in which the number of highly qualified people is divided by the total of technical people (including medical doctors where appropriate).

| Highly qualified / total technicians | Score |
|---|---|
| 0 | 2 |
| 0.01 – 0.19 | 4 |
| 0.20 – 0.39 | 6 |
| 0.40 – 0.49 | 8 |
| 0.50 – 0.59 | 10 |
| >0.60 | 12 |

**Photographic laboratory:** One of the most remarkable advantages of an AS, easy to see and easily detected in the cost-effectiveness right from the beginning, is the substitution of the photographic lab, since through a range of hard-copy systems it is possible immediately, simply by pressing a button, to obtain a copy of a karyotype with photographic quality.

Due to the fact that in Spain, almost always, karyogram photographs are taken (usually for filing purposes), the acquisition of an AS will be most cost effective when there is a dark-room or photographic lab in the cytogenetic unit, since it not only saves time and material, but liberates technicians belonging to the cytogenetic staff. On the contrary, when a photo lab does not exist in the cytogenetic unit, but instead photographs are taken using, generally, the hospital main photo lab (which is more cost effective), AS will save money only.

The score proposed is 2 when there is dark room in the lab, and 1 when not.

**Population:** In order to evaluate the future demand of analysis from the population, we simply consider the type of institution where the laboratory is situated, recognising the fact that labs belonging to a center of a wide population area are liable to support more easily a fast increase in demand; on the contrary, local centers or small labs belonging to small areas usually remain with the same structure, and when demand increases new local labs are founded instead of the old ones being enlarged.

This score goes from 1 for local institutions to 3 for regional or central institutions.

| Center | Score |
|--------|-------|
| Local | 1 |
| County | 2 |
| Regional | 3 |

**Research:** Finally, in those labs where some type of research is undertaken, principally applied research, requiring cytogenetic studies in some methods, or the introduction of new cytogenetic techniques, new applications of the AS could be possible, and, of course, the cost effectiveness would be then better.

| Research | Score |
|----------|-------|
| No | 0 |
| Yes | 2 |

The total quantitative evaluation score is obtained by adding the partial scores of the above items.

## 2.2. Qualitative evaluation

The qualitative evaluation tries to advise about the necessary performance of the automatic system: with or without a metaphase finder, and the speed of the finder if it is required. For this (Figure 2) we evaluate a) the number of samples for considering the need of a finder, and b) the type and number of samples for considering the necessary speed of the finder. The evaluation is performed as follows:

a) After the previous evaluation (see above), we chose 500 as the number of samples per year from which an AS with an automatic metaphase finder will be cost effective. Therefore, in any lab with a throughput of less than 500 samples/year an AS with only automatic karyotyping would be advisable.

b) Due to the differences between finders, not only in the speed of scanning slides but in prices (the price increase progresses with speed), we computed a formula which considers not only the number of cases but the type of

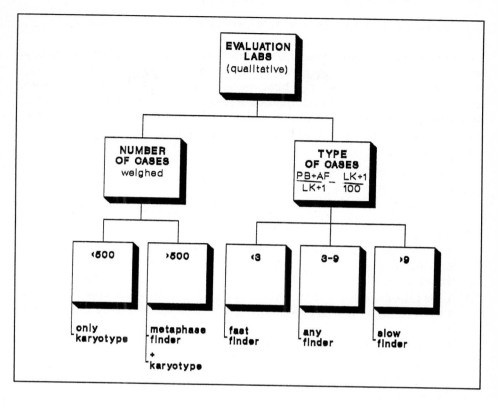

**Fig. 2.** Qualitative laboratory evaluation procedure

samples, referring principally to leukemia (LK) samples as different to AF or PB samples. This formula,

$$\frac{PB + AF}{LK + 1} - \frac{LK + 1}{100}$$

(see Figure 2) arises for the following reasons: 1) there is a physical limitation of 8 or 16 slides per microscope stage, which means from 8 to a maximum of 16 slides/day can be studied by slow finders; 2) usually the cultures from PB and AF are good enough to analyse one case with a single slide, but on the contrary, samples of LK or other cancer tissues usually have to be studied on several slides, not only because of the poorer culture quality, but also in order to analysing different cellular clones. The qualitative classification with the figures obtained from this formula is:

$$<3........ \quad \text{fast finder}$$
$$3 - 9....... \quad \text{any type of finder}$$
$$>9........ \quad \text{slow finder}$$

## 3. Evaluation of automatic systems

Obviously, the evaluation of AS will be considered prior to the acquisition when this is decided. As mentioned above, it would be interesting if a test is widely accepted internationally. Such a test should be directed to the potential users, supplying them with information which helps them to choose the system type and performance most suited to their own characteristics. We think that this test has to consider, at least, the following points:

1) Hardware:
    1.1) Image acquisition
    1.2) Image processing
    1.3) Screen definition
    1.4) Hard copy quality
2) Software:
    2.1) Processes carried out
    2.2) Ease of use
    2.3) Updates
3) User interaction:
    3.1) Number of interactions
    3.2) Interaction devices
    3.3) Easiness to use, time of tiredness
    3.4) Easiness to train
    3.5) Qualification required
4) Type of sample suited to automated analysis, and timings:
    4.1) General:
        4.1.1) Peripheral blood
        4.1.2) Amniotic fluid
        4.1.3) CVs
        4.1.4) Bone marrow
    4.2) Other:
        4.2.1) Aberration scoring
        4.2.2) SCE scoring
        4.2.3) In situ hybridization
5) Maintenance, cost, and time in repairs.
6) Update facilities.
7) Facilities to link to other AS or databases.

## 4. Data

In order to get an evaluation of Spanish cytogenetic laboratories which will serve, firstly, to validate the proposed methodology, we sent a questionnaire to the labs about their structure and characteristics.

**Table 1.** Spanish laboratories: Structure

|  | Mean | Range |
|---|---|---|
| No. of studies | 797 | 317 – 1508 |
| No. of studies, weighted | 938 | 339 – 1844 |
| Staff: |  |  |
| Total | 8.1 | 4 – 13 |
| Medical people* | 2.5 | 1 – 7 |
| Tech. + Aux. | 4.5 | 2 – 8 |
| Medical-tech. + Tech. | 5.0 | 2 – 10 |
| Turnaround times: |  |  |
| Blood | 35.6 | 15 – 90 |
| Prenatal diagnosis | 19.5 | 14 – 30 |

*Only in 2 labs without technical work

**Table 2.** Spanish laboratories: Number of samples, and photo labs

|  | No. labs |
|---|---|
| Number of studies: |  |
| <500 | 3 |
| 500 – 1000 | 7 |
| 1000 – 1500 | – |
| >1500 | 4 |
| Type of studies: |  |
| Blood | 1 |
| Blood + BM | 3 |
| Blood + AF | 1 |
| Blood + BM + AF | 9 |
| Photographic lab: |  |
| Yes | 9 |
| No | 5 |

# 5. Results

At the time of writing we have obtained 14 answers to the questionnaire sent to the Spanish cytogenetic laboratories.

The structure and characteristics of the labs are shown in table 1 (number of samples/year, staff composition, and turnaround times). The type of samples carried out by these labs are shown in table 2.

Evaluating these labs in accordance with the method proposed, the scores in both classifications (quantitative and qualitative) are shown in Table 3.

**Table 3.** Evaluation of 14 Spanish cytogenetic laboratories

| Quantitative | | Qualitative | |
|---|---|---|---|
| score | No. labs | performance | No. labs |
| | | KT only | 3 |
| <10 | 1 | MF + KT | 11 |
| 10 – 15 | 3 | | |
| 15 – 20 | 5 | fast finder | 2 |
| 20 – 25 | 2 | any finder | 4 |
| >25 | 3 | slow finder | 8 |

Figure 3 shows the distribution of the performance of automatic systems from qualitative evaluation (in bars) within the general distribution of quantitative score (area).

We point out that the low quantitative scores (left side of area, Figure 3) are associated with labs in which an AS with karyotyping only (KT) capability is advisable, while the higher quantitative scores point to the advisability of an AS with a fast metaphase finder in addition to automatic karyotyping (MF+KT).

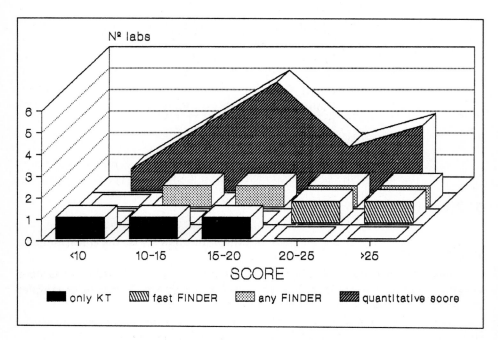

**Fig. 3.** Spanish laboraties evaluation

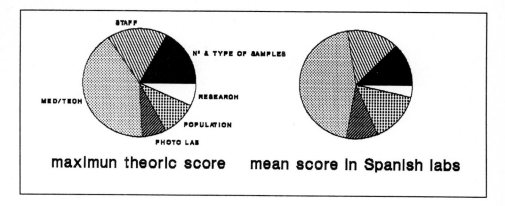

**Fig. 4.** Quantitative evaluation of Spanish laboratories

## 6. Discussion and conclusions

The results of evaluation carried out in the 14 Spanish cytogenetic labs show the possible efficiency of the whole methodology proposed since the lower quantitative scores obtained fit with the simplest performance of an automatic system obtained with the qualitative evaluation, and vice versa.

On the other hand, the mean of the scores carried out in the different items had a distribution (Figure 4) into the total score similar to the theoretical distribution using the maximun possible partial scores.

Finally, we believe that this approach of obtaining objective figures when considering the characteristics of cytogenetic labs, will serve to determine the cost effectiveness of AS and to programme their introduction when it is planned globally, and can also serve as an indicator of the efficiency of individual laboratories.

*Acknowledgements.* I wish to thank C. Lundsteen and J. Piper for their kind suggestions and comments on this paper. This work was supported by grant of Fondo de Investigaciones Sanitarias de España FISS 87/834.

## References

1.    Lubs HA, Priest J, Dev VG, Donahue R, Urbano R, Carpenter A. The effect of automated chromosome analysis on laboratory effieciency: Baseline data from thirty american cytogenetics laboratories. Proc. 8th European Workshop on Automated Cytogenetics, Berlin (1986)

# Cost Evaluation of Magiscan 2

Michael G. Daker

## Summary

Data is presented to show that under the conditions of staff structure and work practices that exist in this laboratory, it takes the same time to complete a full diagnostic chromosome investigation using the Magiscan 2 system as it does using conventional microscopy. This is discussed in relation to cost effectiveness.

## 1. Introduction

The Magiscan 2 chromosome analysis system has undergone extensive clinical trials in Copenhagen (Philip and Lundsteen, 1985; Lundsteen et al., 1986; Lundsteen et al., 1987). The same system has also been evaluated by the Chicago group (Martin, 1985). Both have reported that substantial increases in work load may be achieved using the semi-automated system of chromosome analysis that Magiscan offers.

In 1984, the Department of Health and Social Security (National Health Service Procurement Directorate) initiated an independent evaluation in the United Kingdom, which was carried out in the Cytogenetics Laboratory at Guy's Hospital. In terms of staff structure and operational policies, this laboratory can be considered as typical of the majority of cytogenetics centres in the UK. A full account of the evaluation is given in a DHSS report (Daker, 1988); in this paper, consideration is given to the costs involved in using Magiscan for routine diagnostic chromosome analysis.

## 2. Materials and methods

The system under evaluation, Magiscan 2, was manufactured by Joyce Loebl of Gateshead, UK.[1] This is a semi-automated, single user, interactive system for metaphase finding, chromosome counting, and karyotyping.

The hardware consisted of the following components; (i) the Magiscan 2 unit, with integral monochrome display screen, two floppy disc drives, light-pen and keyboard; (ii) Leitz Orthoplan microscope with 16× and 100× objectives with inverting zoom attachment. The microscope was fitted with a Märzhäuser

---

[1] Magiscan 2 has now been superseded by a more recent model incorporating a number of significant improvements.

**Automation of Cytogenetics**   Editors: C. Lundsteen  J. Piper
© Springer-Verlag Berlin Heidelberg New York 1989

stage which held a maximum of eight slides. The stage movements, zoom and focussing were all motorized; (iii) Microscope control unit; (iv) Bosch video camera; (v) Honeywell VGR4000 video copier, using dry silver type paper; (vi) Epson FX80 dot matrix printer.

The software used in the evaluation provided programmes for metaphase finding, counting and karyotyping using G-banded chromosomes.

The system was used routinely for a period of about twelve months for diagnostic work involving studies of G-banded chromosomes from both blood and amniotic fluid cell (AFC) cultures. At the same time, data was also accumulated from work carried out using conventional microscopy. In order that valid comparisons could be made between the two methods of chromosome analysis, only those cases that conformed to the following criteria were included in the evaluation:-

(i) that a standard investigation had been carried out consisting of 5 spreads analysed in depth (for conventional microscopy this was by direct observation through the microscope: karyotypes were not prepared), together with an additional 6 spreads counted, giving a total of 11 cells examined for a complete investigation,

(ii) that a normal chromosome complement (i.e. 46,XX or 46,XY) had been found, all abnormal results being excluded from this part of the evaluation.

## 3. Results

### 3.1. Data collection

Extensive data was collected on most aspects of the Magiscan system, including timings for chromosome counting and karyotyping. Similar information was also recorded while carrying out chromosome investigations using conventional microscopy (Tables 1 and 2). Using pooled data from five different Magiscan operators (their cytogenetic experience varying from two months to four years), the mean calculated[2] case time for blood chromosome studies was approximately one hour, with a range of 50 to 80 minutes (Table 1). The mean calculated case time for amniotic fluid cell (AFC) cultures was approximately 47 minutes. (Significantly lower, since it was based on the data of only two workers who were both experienced cytogeneticists.) When case times for a single worker were compared, there was little difference between the means for blood and AFC studies (Table 2).

The most significant finding was the fact that the time taken to complete a full cytogenetic study using Magiscan (excluding metaphase finding time, since the scanning system is normally run overnight) was essentially equal to the time taken to carry out the same investigation using conventional microscopy.

---

[2] i.e. the sum of 6 × mean count time and 5 × mean analysis time, plus setting up time 3.5 minutes and metaphase review time 8.0 minutes

**Table 1.** Timing data using the Magiscan system - blood chromosome investigation (data based on the last five cases each of five Magiscan users. The cytogenetic experience of these individuals ranged from 2 months to 4 years).

| Mean time to count one cell | Range of observers' means | Mean time to karyotype one cell | Range of observers' means |
|---|---|---|---|
| 57 s | 34s – 77s | 8m 33 s | 6m 21s – 12m 01s |
| Calculated complete case time[2]: | | 62m (range: 50m – 80m) | |

**Table 2.** Magiscan and conventional analysis - comparison of mean overall case times (with ranges) for a single observer (N.B. data gives mean of actual case times as opposed to the "calculated" case times in Table 1).

| | Magiscan | | Conventional | |
|---|---|---|---|---|
| | No. of cases | mean time per case (mins) | No. of cases | mean time per case (mins) |
| Blood chromosome investigations | 8 | 53 (range 30 – 60) | 12 | 48 (range 24 – 70) |
| AFC chromosome investigations | 15 | 45 (range 30 – 60) | 8 | 60 (range 38 – 83) |
| Combining blood and AFC data | | 49 | | 54 |

## 3.2. Cost estimates

Comparative estimates were made of the costs involved in the analysis of 325 prenatal specimens per year using both the Magiscan system and conventional microscopy. The data is given in Tables 3 and 4.

The figure of 325 prenatal specimens per year was taken as the average annual work load for a trained cytogeneticist. This was based on information collected by a working party of the Association of Clinical Cytogeneticists (ACC) in 1984. The data being derived from a number of laboratories of differing size, within the UK. The work loads of our own laboratory agree with this figure.

Other assumptions that were made are as follows:-

(i) that a formal karyotype (i.e. conventional photographic karyotype or computer generated hard copy) would not be required for diagnostic purposes (N.B. very few laboratories within the UK prepare karyotypes on a routine basis);

(ii) that the system would be used exclusively for AFC chromosomes, on the grounds that automation is more suitable for AFC chromosomes than

**Table 3.** Estimate of costs for 325 AFC specimens using - conventional microscopy (culture costs excluded)

| Staff salary (one wholetime equivalent) (including superannuation and national insurance) | | | £ 11,000 |
|---|---|---|---|
| Equipment | Basic cost | Cost/year | |
| Microscope (assuming 10 year life) | £ 5,000 | £ 500 | |
| Maintenance | | £ 100 | |
| | | £ 600  +VAT | £ 690 |
| | | TOTAL | £ 11,690 |

**Table 4.** Estimate of cost for 325 AFC cases per worker per year using the Magiscan system (Culture costs excluded.) (N.B. 2 workers per system = a total of 650 cases per year per system)

| Staff salary (two wholetime equivalents) (including superannuation and national insurance) | | | £ 22,000 |
|---|---|---|---|
| Equipment | Basic cost | Cost/year | |
| Microscope (assuming 10 year life) | £ 16,540 | £ 1,654 | |
| Magiscan system (assuming 7 year life) | £ 54,945 | £ 7,849 | |
| Installation charge | £ 400 | £ 57 | |
| Maintenance | | £ 7,330 | |
| | | £ 16,890  +VAT | £ 19,424 |
| | | TOTAL | £ 41,424 |
| Since one system will handle 650 cases per annum, the cost of 325 cases per annum will be | | | £ 20,712 |

blood chromosomes;

(iii) that the work regime adopted would enable the system to be used for the maximum number of hours during the normal working day;

(iv) that the staff involved in this work would all be graduate Scientific Officers at the level of Basic Grade.

**Table 5.** Costs for 325 AFC cases including karyotype production – conventional microscopy

| | | |
|---|---|---|
| Culture and analysis costs | | £ 11,690 |
| (salary and microscope, etc.) | | |
| Photographic costs: | materials | £ 1,044 |
| | salary | £ 5,500 |
| | TOTAL | £ 18,234 |

**Table 6.** Costs for 325 AFC cases including karyotype production – Magiscan system

| | |
|---|---|
| Culture and analysis costs | £ 20,712 |
| (salary and Magiscan system) | |
| Hardcopy (copy unit, maintenance, and dry silver paper) | £ 1,672 |
| TOTAL | £ 22,384 |

Further cost estimates were then prepared based on the assumption that karyotyping would be essential. The costings were recalculated as follows in tables 5 and 6, assuming that two karyotypes would be prepared for each case, and that each karyotype would take one hour to prepare.

# 4. Discussion

Using the Magiscan system, the mean case time for this laboratory was observed to be approximately one hour. Although this appears to be substantially higher that the figure of 35 minutes quoted by Lundsteen et al., (1987), the difference is due very largely to variations between the laboratories in the number of cells karyotyped, and possibly also to variation between Magiscan operators. Table 1 shows a relatively wide range of mean case times for the five workers involved. In general, the faster times were recorded by the more experienced workers, but there were exceptions to this, suggesting that some individuals showed greater aptitude that others for using a computer operated system. Philip and Lundsteen (1985) report that substantial increases to laboratory workload may be achieved by the introduction of automation. Thus, for a laboratory currently making 1,000 analyses a year, and employing four technicians, it is suggested that the capacity could be doubled by the introduction of one semi-automated system. Martin (1985), also using the Magiscan system, reports a saving of at least one hour per case, which she equates to half a technician salary per year. This, in turn, would allow the processing of an additional 100 patients per year, to give a machine pay back time of approximately three years.

Our own experience was less encouraging. Central to our own calculations of cost effectiveness was the observation that the time taken to complete a full examination of G-banded chromosomes of a patient using the Magiscan system was virtually the same as when using conventional microscopy. Since there was no saving in time, and since hardcopy karyotypes were not required, it was clearly not possible to operate Magiscan cost effectively using the approach that we adopted for the evaluation (i.e. using the Magiscan system as a direct equivalent of a conventional microscope).

It is difficult to draw comparisons between laboratories, in view of the differences that exist in work practices, staffing levels, and the grade of staff, etc., but it is a matter of considerable concern that our findings should be so much at variance with those of others. It had always been anticipated that any difference between our results and those of the workers in Copenhagen and Chicago would be explained simply by the fact that in the UK a hard copy karyotype is generally not required. However, the data in Tables 5 and 6 do not support this.

A particular feature of our evaluation was the way in which data was collected for the timing studies, whereby the only cases included were those with a normal chromosome complement, and where the cytogeneticist had adhered rigidly to the standard regime of five spreads analyzed; six spreads counted. The mean case times observed, therefore, represent minimal values. It has to be remembered that throughout the year, the total work load may include cases of mosaicism, marker chromosomes, fragile site studies, and structural rearrangements that may be difficult to interpret, as well as high resolution banding studies and special staining techniques. All these may require substantially longer periods of analysis. The ACC figure of 325 cases per cytogeneticist per year serves to emphasize this point. Assuming that a working year consists of 220 days, and a working day, 375 minutes (ACC Working Party figures), then on average each case takes over four hours. A proportion of this will be culture time (approximately 50% for AFC specimens), but a large proportion of the remainder will be taken up by analysis.

These arguments explain the apparently low number of cases that we have proposed could be analyzed in a year using one Magiscan system. At the same time, they probably also account for some of the differences between our own conclusions and those of other groups. Certainly, we feel that it is important to realize that the case times observed in our evaluation represent minimal values, and should not be used for calculating theoretical work loads.

In reality, the difference between the total costs shown in Tables 5 and 6 is relatively small, and it is not difficult to visualize changes in work practice that could make automation more cost effective. For example, in the UK virtually all cytogenetic staff are university graduates; using less well qualified staff to provide the karyotypes, from which the experienced cytogeneticists could then make their diagnoses, would almost certainly make the system a viable proposition.

For those laboratories where hard copy karyotypes are considered essential, Magiscan could be used to follow an entirely different regime to achieve greater cost effectiveness. Thus, if a complete system were to be dedicated solely to karyotyping cells already analyzed by conventional microscopy, then the cost effectiveness of the system would rise rapidly as the workload increased. In the UK, the largest laboratories have an AFC work load of about 2,000 cases per year; it can be calculated that the use of Magiscan in this situation might save about one half of a salary per year. However, laboratories handling four or five thousand cases a year would stand to make very substantial savings.

The evaluation programme that was followed required a straightforward assessment of the Magiscan system for metaphase finding, chromosome counting and analysis. There are, however, other aspects of the system that now enter the equation. In particular, the chromosome straightening software is of great potential value, and must surely become the method of choice for handling extended chromosomes. How this will effect cost analysis calculations is difficult to assess, and indeed, perhaps it is wrong to attempt to justify this sort of system purely on the grounds of cost effectiveness. Computer aided chromosome analysis may have "arrived", but what is required now is widespread application of these systems so that their potential can be developed to the full.

*Acknowledgements.* I would like to thank the DHSS for setting up the evaluation, and for their valuable support throughout. I am also grateful to Joyce-Loebl for their friendly assistance at all times. My thanks also to Lynne Noble for her help in the preparation of this manuscript.

# References

1    Daker MG. An evaluation of the Joyce-Loebl Magiscan 2 chromosome analysis system. DHSS Report number STD/88/15 (1988)
2    Lundsteen C, Gerdes T, Maahr J. Automatic classification of chromosomes as part of a routine system for clinical analysis. Cytometry 7:1-7 (1986)
3    Lundsteen C, Gerdes T, Maahr J, Philip J. Clinical performance of a system for semi-automated chromosome analysis. Am. J. Hum. Genet. 41:493-502 (1987)
4    Martin AO. My life with two automated systems. In: Preliminary report of work group on automated chromosome analysis. Supported by the EEC. Leiden (1985)
5    Philip J, Lundsteen C. Semi-automated chromosome analysis. A clinical test. Clin. Genet. 27:140-146 (1985)

# Part III

# Flow Cytogenetics

# The Flow Cytometry Approach to Automated Chromosome Analysis

Judith A. Fantes, Daryll K. Green

## 1. Introduction

In contrast to the detailed morphological study of chromosomes on a microscope slide the information obtained from analysis of chromosomes in suspension by flow cytometry is of a more statistical nature, resulting from measurement of many thousands of chromosomes in a short time. It can be seen for example in the chapter by Alexander Cook that a small deletion or addition of DNA to a particular chromosome of an individuals karyotype, which may be invisible to microscope analysis, produces a significant shift in the mean expected position of the fluorescence peak of that chromosome in the flow karyotype (Harris *et al* 1987; Gray *et al* 1988). Typically, this result could be obtained from a suspension of $10^5$ chromosomes in 60 seconds. Most types of homogeneous disturbance to the distribution of DNA in the karyotype, which are greater than the average size of a chromosome band (600 band set) could be observed by flow cytogenetics (Green *et al* 1984a) and this information complements the detailed cytogenetic and molecular information provided by microscopic examination and DNA probe hybridization (Wilcox *et al* 1986).

A number of research groups are also investigating the potential of flow cytogenetics to measure heterogeneous changes in the karyotype, particularly those induced by clastogens (Aten *et al* 1984; Green *et al* 1984b; Cremer *et al* 1982; Otto and Oldiges 1980). Some advances have been made in this area and exposures greater than 0.5G have been detected by flow cytometry of metaphase chromosomes (Green *et al* 1984b). Unfortunately at the moment these methods are not sensitive enough to detect doses just above background, and this must still be monitored by time consuming, labour intensive methods of aberration analysis of metaphase cells. Approaches where flow cytogenetics could make an important contribution to automatic dosimetry will be discussed here.

The most significant contribution which flow cytogenetics has made to the complete knowledge of chromosomes has been the production of enriched chromosome samples. Chromosomes from a selected group, sorted by flow into a 90-95% enriched suspension of that chromosome have provided the molecular biologist with a powerful tool to look beyond the metaphase chromosome into

**Automation of Cytogenetics**   Editors: C. Lundsteen J. Piper
© Springer-Verlag Berlin Heidelberg New York 1989

the detailed DNA composition (Deaven *et al* 1986). A dictionary of chromosome information consisting of positive identification by microscopy, accurate DNA measurement by flow and a map of DNA markers from the molecular approach is leading to accurate links between hereditary diseases and identifiable genes. Flow cytogenetics has an important part to play in this process.

## 2. Preparation of Chromosomes for Flow Analysis

Isolated chromosomes can be prepared from cultures of peripheral blood lymphocytes, EBV transformed lymphoblastoid cells and fibroblasts cells, which have been blocked at metaphase using a mitotic inhibitor such as colchicine. The cells are then swollen in a hypotonic solution to separate the chromosomes and the cell walls are broken by detergent action followed by mechanical disruption. The chromosomes are released into a buffer solution where they are stabilised by polyamines (Sillar and Young, 1981), magnesium (Van den Engh *et al*, 1985) or by intercalating dyes (Aten *et al*, 1984). Chromosomes stabilised by polyamines are highly condensed and produce well resolved flow karyotypes suitable for detection of abnormalities and sorting. For slit scanning, techniques which result in elongated chromosomes are neccessary (Lucas and Gray, 1987), and for antibody labelling buffers are used which retain the antigenicity of the chromosomes (Trask *et al*, 1984).

## 3. Dual Beam Sorting

A suspension of metaphase chromosomes when stained with a single DNA specific fluorescent dye and measured at speeds of 1000/sec. by single beam flow cytometry produce a fluorescence intensity distribution quite unique to the donor (Green *et al* 1986). As many as 16 resolvable peaks can appear in the distribution (flow karyotype), some of which correspond to single chromosome groups and which result, after sorting, in >90% pure enriched fractions of individual homologues or sex chromosomes. Dyes such as ethidium bromide or propidium iodide, which intercalate into the DNA produce one kind of pattern useful for some chromosome groups (e.g. 3,4,21 and 22) while dyes such as Hoechst 33258 and Dapi produce a different pattern useful for other chromosome groups (e.g. Y,13,14 and 15). The latter two dyes bind preferentially to AT rich DNA and are excited by UV light and when combined with another dye Chromomycin A3, which binds preferentially to GC rich DNA and is excited by blue light, a further increase in chromosome peak resolution in the two colour karyotype is obtained. An example of Hoechst/Chromomycin fluorescence distribution of male human chromosomes harvested from a lymphoblastoid cell line is shown in Figure 1. There are 20 peaks which are particularly well resolved for the small chromosome groups. Many laboratories have strived to improve the resolution of the two colour distribution, usually by adjustments

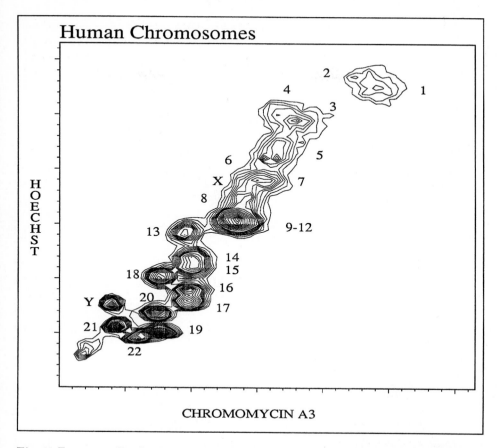

**Fig. 1.** Frequency distribution of Hoechst and Chromomycin A3 fluorescence intensity of human chromosomes, shown in contour presentation. Identification of the peaks was established by sorting chromosomes giving rise to particular peaks followed by analysis of the sorted samples under a light microscope. Flow karyotypes similar to that above were first published by Gray *et al* in 1979.

to the preparation of the chromosome suspension (van den Engh *et al* 1988) and they have recorded very high purities for sorted samples of all but the 9-12 group chromosomes.

Chromosome material sorted in this way can be used to construct specific DNA libraries (Krumlauf *et al* 1982) or can be sorted onto filter papers for gene mapping to particular chromosomes (Lebo *et al* 1984) or recently as a source of material for pulse field gel electrophoresis and long range mapping.

DNA libraries are normally constructed from approximately 2 million sorted chromosomes and a sample of this size sorted on a conventional type of flow sorter would be accumulated after 3 to 4 working days of sorting. Successful attempts have been made to significantly reduce the sorting time by

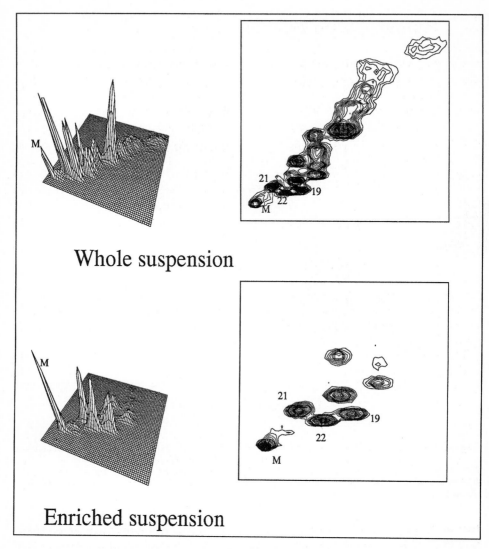

Whole suspension

Enriched suspension

**Fig. 2.** At top right a contour plot of the Hoechst/Chromomycin A3 fluorescence distribution of all the chromosomes in a normally prepared suspension is shown. A small marker chromosome (labelled "M") present in all cells, which is thought to be derived from a 22 chromosome, forms a peak at the bottom left corner, and in the adjacent (top left) isometric plot of the same chromosome distribution the proportion of the marker chromosome can be clearly seen to be relatively small. An expanded view of the small chromosome distribution of a gradient enriched suspension is shown at the bottom right and left. No large chromosomes were present in the suspension and clearly the proportion of the marker chromosome has been greatly increased.

construction of a high speed sorter (Gray *et al* 1987; Peters *et al* 1985) operating at an order of magnitude greater speed than is normal. This is however an expensive and technically difficult solution. We have adopted a simpler and cheaper approach, which works well for the smaller chromosomes and involves chromosome enrichment through a glycerol gradient before flow sorting. Figure 2 shows a comparison by contour and isometric plot of the proportion of a small abnormal marker chromosome (1/3 size of chromosome 22) in a standard and gradient enriched suspension analysed with dual beam flow cytometry. An order of magnitude enrichment is quite normal although the overall density of chromosomes in the enriched sample is less than the original unenriched sample. Gradient enrichment can improve the sorting speed of a particular chromosome by a factor of 4 or 5 and in many cases with this approach one working day's sorting produces the 2 million chromosomes required for DNA library construction.

# 4. Fluorescence through Antibodies

An important and expanding area of chromosome analysis by flow technology is developing from the use of antibodies against chromosomal proteins. One study involved incubating a chromosome suspension with a monoclonal antibody to the chromosomal protein, histone 2B, followed by a second FITC conjugated anti mouse antibody (Trask *et al* 1984). Binding of the antibody was proportional to DNA content, so no additional karyotype information was gained but the antibody is now being used as a probe in flow cytometry studies of chromosomal structure (Turner and Keohane 1987).

One particular area which has received a good deal of attention by several researchers including the authors, is that of kinetochore labelling by CREST antibodies. The technique was first established on microscope slide preparations (Moroi *et al* 1980) where it was found that antibodies in the sera of patients with the CREST variant of the disease scleroderma bound specifically to protein antigens at the kinetochore region of the chromosome. We have developed methods using these sera for labelling the kinetochores of chromosomes in suspension (Fantes *et al* 1988). The chromosomes are prepared in a buffer which retains the antigenicity of the kinetochore proteins and then incubated with CREST serum before centrifuging through a glycerol gradient to remove excess serum while reducing chromosome aggregation. Finally the chromosomes are incubated with FITC conjugated rabbit anti human IgG to give a fluorescence signal at the centromere, counterstained with a DNA stain such as Hoechst 33258 or DAPI, and analysed by flow. A typical two colour distribution is shown in Figure 3, which shows all the features of a DNA flow karyotype together with a reasonably uniform distribution of centromere fluorescence intensity. There is a slight but significant correlation of FITC staining with chromosome size; most of the CREST sera also react with a chromosome

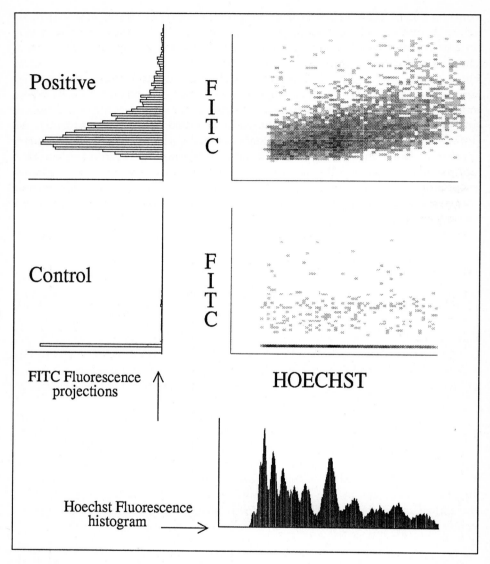

**Fig. 3.** DNA bound Hoechst and CREST bound FITC fluorescence distributions are shown for CREST positive and control samples. In each case a scattergram is shown on the right and a condensation of the FITC fluorescence distribution is shown on the left. At the bottom right is shown a condensation of the Hoechst fluorescence distribution for either positive or control samples.

associated but not centromeric antigen (Jeppeson and Nichol 1986) and this leads to binding along chromosome arms. Figure 3 also shows the result of a control sample using serum from a normal female where there is no measurable labelling of the centromeres.

The ultimate aim of this particular study was to detect and count dicentric chromosomes (chromosomes with two centromeres) in flow. Dicentric chromosomes together with acentric fragments are the most commonly occurring unstable chromosome aberrations induced by ionizing radiation and their presence in peripheral blood lymphocytes is used as an accurate measure of radiation exposure (Evans and Lloyd 1978). Eventually it may be possible to detect dicentric chromosomes in flow using the methods described above as they have should have twice the fluorescence of a monocentric chromosome but many problems such as chromosome aggregation, loss of large chromosomes during preparation and low signal to noise of the immunofluorescence must be resolved before this technique can sensibly be used to monitor radiation dose.

## 5. Centromeric Index

Other methods for detecting the presence and position of chromosome centromeres have been developed from slit-scan flow cytometry. Here the fluorescence from chromosomes whose long axes are along the direction of flow, is measured with a narrow aperture, typically about $1\mu$m wide. A profile of fluorescence along each chromosome is recorded and a clear dip in the profile denotes the position of the centromere allowing a centromeric index to be calculated (Lucas et al 1987). Because of the large amounts of signal processing required for each chromosome the flow rate of these instruments falls well below the normal 1000 or more per second; 50 chromosomes/sec is typical; but the measurement of centromeric index has opened the way for separating the 9-12 group chromosomes which have a similar DNA content and AT/GC base pair ratios.

Dicentric chromosomes can also be detected by the slit-scan approach. Particular care has to be taken to distinguish true dicentric chromosomes from clusters of monocentric chromosomes (Lucas et al 1987) and because of stringent selection criteria many true positives are missed. This technique is not yet fast enough or straightforward enough to challenge manual chromosome analysis for dosimetry measurement.

## 6. Future Advances

Technical advances in slit-scanning flow cytometry will mean that the remaining chromosomes which are not easily analysed and sorted by fluorescence intensity measurements, notably the 9-12 group chromosomes, will become separable and sorted at realistic speeds by centromeric index measurements. This also applies to marker chromosomes with similar amounts of AT and GC rich DNA to normal chromosomes but with differing arm lengths. Slit-scanning CREST antibody stained chromosomes may also lead to more accurate counting of mono and dicentric chromosomes for radiation dosimetry.

Recently it has become possible to unambiguously identify individual chromosomes in a metaphase cell by *in situ* hybridisation using cloned DNA probes detecting highly repeated sequences (Yang *et al* 1982). At the same time techniques of non radioactive *in situ* have been developed using biotinylated DNA probes and detecting the sites of hybridisation by antibodies conjugated to enzymes or fluorescent dyes (Pinkel *et al* 1986), so that it is now possible to pick out a specific chromosome by fluorescence. This is now being applied to chromosomes in suspension, (Dudin *et al* 1987), and although there are problems of aggregation it should eventually be possible to analyse and sort chromosomes by their DNA signature rather than DNA content.

# References

1.  Aten JA, Kooi MW, Bijman JTh, Kipp JBA, Barensden GW. Flow cytometric analysis of chromosome damage after irradiation: Relation to chromosome aberrations and cell survival. In Biological Dosimetry, Eds. Eisert WG, Mendelsohn ML. pp57-60, Springer-Verlag, 1984.
2.  Cremer C, Cremer T, Gray JW. Induction of chromosome damage by ultra violet light and caffeine: Correlation of cytogenetic evaluation and flow karyotype. Cytometry 2, 287-290, 1982.
3.  Deaven LL, Van-Dilla MA, Bartholdi MF, Carrano AV, Cram LS, Fuscoe JC, Gray JW, Hildebrand CE, Moyzis RK, Perlman J. Construction of human chromosome-specific DNA libraries from flow-sorted chromosomes. Cold Spring Harbor Symp.Quant.Biol. 51, 159-167, 1986.
4.  Dudin G, Cremer T, Schardin M, Hausmann M, Bier F, Cremer C. A method for nucleic acid hybridization to isolated chromosomes in suspension. Human Genet. 76, 290-292, 1987.
5.  Evans HJ, Lloyd DC. Editors, Mutagen induced chromosome damage in man. Edinburgh Univ. Press, 1978.
6.  Fantes JA, Green DK, Malloy P, Sumner AT. Flow cytometry measurements of human chromosome kinetochore labelling. Submitted to Cytometry, 1988.
7.  Gray JW, Dean PN, Fuscoe JC, Peters DC, Trask BJ, van-den-Engh GJ, Van Dilla MA. High-speed chromosome sorting. Science 238, 323-329, 1987.
8.  Gray JW, Langlois RG, Carrano AV, Van Dilla MA. High resolution chromosome analysis: One and two parameter flow cytometry. Chromosoma 73, 9-27, 1979.
9.  Gray JW, Trask B, van-den-Engh G, Silva A, Lozes C, Grell S, Schonberg S, Yu LC, Golbus MS: Application of flow karyotyping in prenatal detection of chromosome aberrations. Am.J.Hum.Genet. 42, 49-59, 1988.
10. Green DK, Fantes JA, Buckton KE, Elder JK, Malloy P, Carothers A, Evans HJ. Karyotyping and identification of human chromosome polymorphisms by single fluorochrome flow cytometry. Human Genet. 66, 143-146, 1984a.
11. Green DK, Fantes JA, Spowart G. Radiation dosimetry using the methods of flow cytogenetics. In Biological Dosimetry, Eds. Eisert WG, Mendelsohn ML. pp67-76, Springer-Verlag, 1984b.
12. Green DK, Fantes JA, Evans HJ. Human metaphase chromosomes: Analysis and sorting by flow cytometry. In Genetic Disorders of the Fetus, 2nd. Edition, Ed. Milunsky A., pp741-754, Plenum Press, 1986.

13.  Harris P, Cook A, Boyd E, Young BD, Fergusson-Smith MA. The potential of family flow karyotyping for the detection of chromosome abnormalities. Human Genetic. 76, 129-133, 1987.
14.  Jeppeson P, Nicol L. Non-kinetochore directed autoantibodies in Scleroderma/CREST. Mol. Biol. Med. 3, 369-384, 1986.
15.  Krumlauf R, Jeanpierre M, Young BD. Construction and characterisation of genomic libraries from specific human chromosomes. Proc. Nat. Acad. Sci. USA 79, 2971-2975, 1982.
16.  Lebo RV, Gorin F, Flitteric RJ, Kao FT, Cheung MC, Bruce BD, Kan YW. High resolution chromosome sorting and DNA spot blot analysis assign McArdle's syndrome to chromosome 11. Science 225, 57-59, 1984.
17.  Lucas JN, Gray JW. Centromeric index versus DNA content flow karyotypes of human chromosomes measured by means of slit-scan flow cytometry. Cytometry 8, 273-279, 1987.
18.  Lucas JN, Lozes C, Mullikin D, Pinkel D, Gray J. Automated quantification of the frequency of aberrant chromosomes in human cells. Cytometry Suppl. 1, p13, 1987.
19.  Moroi Y, Peebles C, Fritzler MJ, Steigerwald J, Tan EM. Autoantibody to centromere (kinetochore) in Scleraderma Sera. Proc. Nat. Acad. Sci. 77, 1627-1631, 1980.
20.  Otto FJ, Oldiges H. Flow cytogenetic studies in chromosomes and whole cells for the detection of clastogenic effects. Cytometry 1, 13-17, 1980.
21.  Peters D, Branscomb E, Dean P, Merrill T, Pinkel D, Van-Dilla M, Gray JW. The LLNL high-speed sorter: design features, operational characteristics, and biological utility. Cytometry 6, 290-301, 1985.
22.  Pinkel D, Straume T, Gray JW. Cytogenetic analysis using quantitative, high-sensitivity, fluorescence hybridization. Proc.Natl.Acad.Sci.USA 83, 2934-2938, 1986.
23.  Sillar R, Young BD. A new method for the preparation of metaphase chromosomes for flow analysis. J. Histochem. Cytochem. 29, 74-78, 1981.
24.  Trask B, van den Engh G, Gray JW, Vanderlaan M, Turner B. Immunofluorescent detection of histone 2B on metaphase chromosomes using flow cytometry. Chromosoma 90, 295-302, 1984.
25.  Turner BM, Keohane A. Antibody labelling and flow cytometric analysis of metaphase chromosomes reveals two discrete structural forms. Chromosoma 95, 263-270, 1987.
26.  Van den Engh GJ, Trask BJ, Gray JW, Langlois RG, Yu L-C. Preparation and bivariate analysis of suspensions of human chromosomes. Cytometry 6, 92-100, 1985.
27.  Van den Engh G, Trask B, Lansdorp P, Gray JW. Improved resolution of flow cytometric measurements of Hoechst and Chromomycin A3 stained human chromosomes after addition of citrate and sulfite. Cytometry 9, 271-274, 1988.
28.  Wilcox DE, Cooke A, Colgan J, Boyd E, Aitken DA, Sinclair L, Glasgow L, Stephenson JBP, Ferguson-Smith MA. Duchenne muscular dystrophy due to familial Xp21 deletion detectable by DNA analysis and flow cytometry. Human Genet. 73, 175-180, 1986.
29.  Yang TP, Hansen SK, Oishi KK, Ryder OA, Hamkalo BA. Characterisation of a cloned repetitive DNA sequence concentrated on the human X chromosome. Proc. Natl. Acad. Sci. (USA) 79, 6593-6597, 1982.

# Flow Analysis of Human Chromosomes

**A. Cooke**

**Summary**

The potential of a single-laser fluorescence activated cell sorter as an analytical aid to conventional cytogenetics is discussed. Flow karyotype analysis is of particular value when all of the members of a family under investigation are available for study. Standardisation of the fluorescence values of chromosome peaks allows comparisons to be made between flow karyotypes from different individuals and may permit the detection of abnormalities at, or just beyond, the limits of resolution of microscopy.

## 1. Introduction

The use of fluorescence-activated cell sorters (FACS) to examine preparations of human chromosomes has expanded significantly since it was first described (Gray et al 1975). Clearly, the most important function of these machines in genetics is their capacity to provide a source of sorted chromosomes for library construction and the mapping of cloned genes and restriction fragment length polymorphisms. This is particularly true of systems employing dual laser flow sorters which can produce distinct separation of many discrete chromosome populations (Collard et al 1984; Lebo et al 1986). FACS instruments can also be used to directly analyse suitably prepared samples of human chromosomes yielding flow karyotypes which are usually amenable to further computer analysis (Young et al 1981). It is this second basic role of the flow cytometer in the environment of a research/routine human genetics department which is the subject of this brief report.

## 2. Materials and methods

### 2.1. Sample Preparation and Principle of Operation of the Flow Cytometer

The first essential for successful analysis of human chromosomes by flow karyotyping is good quality sample preparation. This has been achieved using fibroblasts, phytohemagglutinin stimulated lymphocytes and transformed lymphoblastoid cell lines. Of these, the third cell type may probably prove the most useful as a high mitotic index can regularly be obtained, and the cell

**Automation of Cytogenetics**   Editors: C. Lundsteen J. Piper
© Springer-Verlag Berlin Heidelberg New York 1989

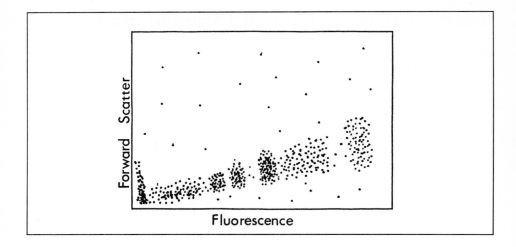

**Fig. 1.** Diagram of dot display pattern of a preparation of human chromosomes.

line easily maintained in culture if a repeat flow analysis is required. Following treatment of an actively growing (and if possible partially synchronised) culture with a suitable spindle inhibitor, the harvested cells are suspended in a hypotonic solution and disrupted mechanically to release the chromosomes. These are then resuspended in an appropriate buffer solution which should have the capacity to stabilise the ᴄhromosomes and prevent nuclease digestion. A good example of this is the polyamine/digitonin buffer described by Sillar and Young (1981). The quality of the sample can be ascertained at this stage by microscopic examination, and if necessary most of the interphase nuclei removed by centrifugation.

At an appropriate time prior to running on the flow cytometer each sample aliquot is stained with a suitable DNA fluorochrome (or fluorochromes). Commonly employed stains include the non-specific ethidium bromide and propidium iodide; A-T specific Hoechst 33258 and DAPI; G-C specific chromomycin A3.

Analysis of the stained chromosomes by the flow cytometer occurs in a fluid stream which is produced by forcing the sample plus carrier buffer through a nozzle which is typically 50-100$\mu$m in diameter. A short distance below the nozzle, the fluid steam is intersected by a laser beam which has been previously set to the wavelength required to excite the fluorescent stain being used. The fluorescence signal is collected by a lens at 90° to the laser beam, and focused on to a photomultiplier tube which converts it into an electrical signal which can then be processed by the instrument. Forward scattered light, giving a measure of the size of the particles passing through the laser beam, is recorded by a second lens in line with the laser beam.

The fluorescence and forward scatter signals for each particle combine to produce a dot display with the former parameter as the horizontal axis, and the latter as the vertical axis. When human chromosomes are analysed by the flow cytometer, they appear on the dot display as an approximately diagonal line originating in the lower left corner of the screen (figure 1), with the size of each chromosome being the principal determinant as to its position in the display.

## 2.2. Flow Karyotype Production and Analysis

Flow karyotypes of human chromosomes are acquired in a single laser system by collecting data from the fluorescence parameter over a short period of time so that a number of particles with identical (or very similar) fluorescence are recorded. The fluorescence axis is divided into a number of channels (up to a maximum number of 1024 on the FACS in our department), and peaks in the flow karyotype are formed by the accumulation of signals in immediately adjacent channels. When a dual laser system of illumination is employed, two fluorescence signals are collected, and a two dimensional chromosome distribution map is produced (van den Engh et al 1985). The flow cytometer in our department has a single 5w argon laser as light source, and it is results obtained with this instrument that will be used to further illustrate some aspects of flow karyotype analysis.

## 3. Results

The quality of resolution of flow karyotypes is as already stated critically dependent on the standard of sample preparation. A second factor which must be considered is the amount of available laser power at the specific wavelength(s) required to excite the particular fluorochrome used to stain the chromosomal DNA. Stains requiring UV wavelengths for excitation such as Hoechst 33258 thus either necessitate the use of more powerful lasers, or require the maximum output from smaller lasers. Provided a suitable system is available, excellently resolved flow karyotypes can be produced with Hoechst 33258 (Green et al 1984; Lalande et al 1984). Prolonged use of a laser at its maximum output is however likely to lead to a significant reduction in lifespan and there are thus obvious advantages for our system in employing a stain such as ethidium bromide which is excited by powerful visible laser lines and has the potential to resolve around 20 separate peaks in the typical individual's flow karyotype (Figure 2).

Flow karyotypes produced with ethidium bromide have been shown to provide an accurate estimate of chromosomal DNA content (Harris et al 1986), and to also resolve a number of individual chromosome homologues (Green et al 1984; Harris 1984). Chromosomes exhibiting the greatest variation in DNA

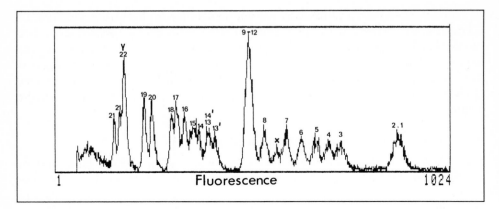

**Fig. 2.** Flow karyotype of normal male stained with ethidium bromide obtained using the 514nm laser line at a power of 0.7w

content are those which have been shown by conventional cytogenetic techniques to have differing amounts of heterochromatin, namely, 1, 9, 16 and Y, followed by the acrocentric chromosomes 13, 14, 15, 21 and 22 (Harris et al 1986). In the example of a normal male flow karyotype shown in Figure 2, it can be seen that the chromosomes 13, 14 and 21 have homologues differing in size. This capacity to clearly resolve normal heteromorphisms could be said to concurrently demonstrate the sensitivity and limitations of flow analysis of human chromosomes. The use of quantitive fluorescent DNA stains can effectively preclude any distinction between an excess (or absence) of heterochromatin or acrocentric satellite material and a possible abnormality in a significant number of chromosomes.

It is however relevant to this point that a high proportion of suspected chromosome abnormalities occur in young children whose parents are also available for investigation. Thus instead of examining the flow karyotype of an individual in isolation, it is potentially much more informative to undertake a family study, involving the child and both parents. Standardisation of the flow karyotypes of family members can be achieved by assigning a given relative fluorescence value (e.g. 560) to the main central peak which contains chromosomes 10, 11, 12 and frequently 9 and calculating all other peak values relative to this using a suitable computer program (Young et al 1981; Harris et al 1986). In this manner it is possible to determine the parental origin of the majority of a child's chromosome homologues which have been resolved by flow analysis, and also to clearly distinguish any *de novo* abnormality or unbalanced translocation which is within the limits of detection of the flow cytometer (Harris 1984; Cooke et al 1987). An example of a family study (in which no abnormality was detected) is shown in Figure 3 and the relative fluorescence values of each

**Table 1.** Relative fluorescence values calculated from the flow karyotypes of the family shown in Figure 3. C = homologue inherited by child; P = paternal; M = maternal.

| FATHER | | CHILD | | MOTHER | |
|---|---|---|---|---|---|
| Relative Fluores. | Chromosome Number | Relative Fluores. | Chromosome Number | Relative Fluores. | Chromosome Number |
| 1020 | 1(x1) | | | | |
| 986 | 1(x1)C, 2 | 988 | 1, 2 | 1009 | 1, 2 |
| 827 | 3 | 819 | 3 | 825 | 3 |
| 789 | 4 | 781 | 4 | 788 | 4 |
| 754 | 5 | 746 | 5 | 755 | 5 |
| 707 | 6 | 703 | 6 | 712 | 6 |
| 664 | 7 | 660 | 7 | 665 | 7 |
| 638 | X | 636 | X | 639 | X |
| 600 | 8 | 601 | 8 | 601 | 8 |
| 560 | 9-12 | 560 | 9-12 | 560 | 9-12 |
| 464 | 13(x1) | | | | |
| 436 | 13(x1)C, 14 | 434 | 13, 14 | 436 | 13,14 |
| 410 | 15 | 403 | 15 | 392 | 15, 16(x1) |
| 377 | 16 | 377 | 16(x1)P | | |
| | | 361 | 16(x1)M | 359 | 16(x1)C |
| 352 | 17 | 352 | 17 | 345 | 17 |
| 336 | 18 | 335 | 18 | 326 | 18 |
| 272 | 20 | 271 | 20 | 275 | 20(x1)C |
| | | | | 264 | 20(x1) |
| 256 | 19 | 254 | 19 | 247 | 19 |
| 216 | Y, 22(x1) | | | | |
| 208 | 22(x1)C | 206 | 22 | 195 | 22 |
| 184 | 21(x1) | 186 | 21(x1)?P | 180 | 21(x1) |
| 175 | 21(x1) | 174 | 21(x1)?M | 171 | 21(x1) |

chromosome peak are given in Table 1. As can be seen from the values in Table 1, the father has passed on his smaller homologue of chromosomes 1, 13 and 22 to the child while the mother has contributed her smaller 16 homologue and larger 20 homologue. Although all members of the family have separately resolved homologues of chromosome 21, the values are too similar to permit confident assignment but it is most probable that the child's larger homologue is of paternal origin and the smaller of maternal origin.

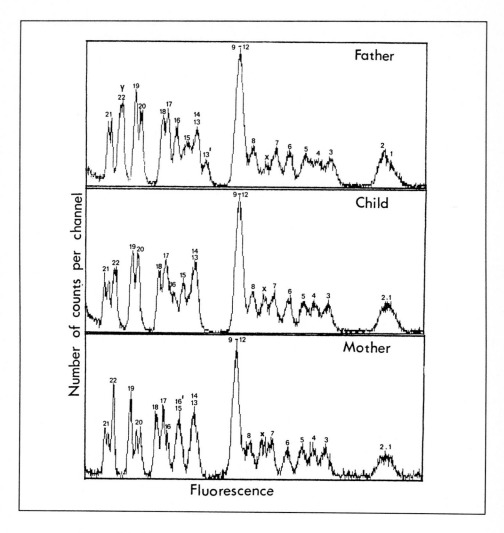

**Fig. 3.** Ethidium bromide flow karyotypes of a child and her parents

Flow karyotype analysis of family members can prove particularly valuable when a small abnormality is suspected, but cannot be definitely confirmed by cytogenetics. In the example shown in Figure 4, the child was diagnosed as having a syndrome known to sometimes involve a deletion of one homologue of chromosome 15 (Angelman syndrome). The presence of a deletion could not be clearly ascertained by microscopy, but detailed flow analysis (Table 2) demonstrated that the child had a *de novo* deletion of 15 and this information could therefore be used in counselling the family.

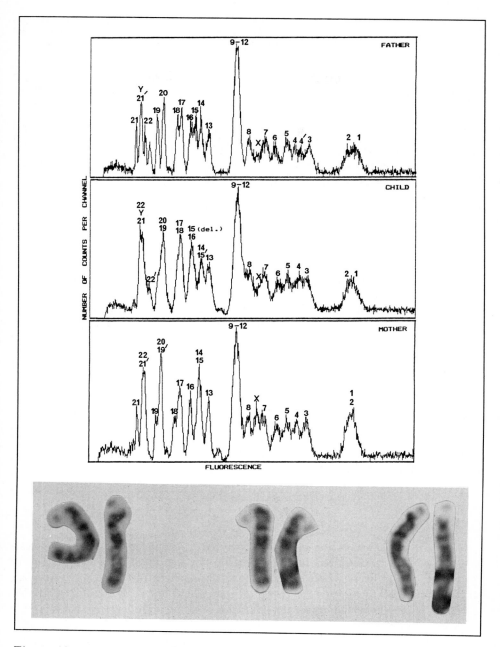

**Fig. 4.** Above – Ethidium bromide flow karyotypes of a child with Angelman syndrome showing a deletion of chromosome 15 and his parents. Below – Chromosome 15's from the affected child.

Table 2. Relative fluorescence values for the child and parents shown in Figure 4. Six flow karyotypes were produced for each individual and the results shown are the mean values and standard deviations. C = homologue inherited by child; M = maternal; P = paternal.

| FATHER | | | CHILD | | | MOTHER | | |
|---|---|---|---|---|---|---|---|---|
| Chromosome Number | Relative Fluores. | | Chromosome Number | Relative Fluores. | | Chromosome Number | Relative Fluores. | |
| | mean | s.d. | | mean | s.d. | | mean | s.d. |
| 1 | 1009.5 | 2.0 | 1 | 1004.9 | 3.3 | 1 | 996.9 | 2.3 |
| 2 | 973.1 | 1.2 | 2 | 973.4 | 3.9 | 2 | 970.2 | 3.1 |
| 3 | 831.1 | 1.5 | 3 | 823.8 | 2.4 | 3 | 819.3 | 1.3 |
| 4(x1)C | 792.0 | 1.5 | 4 | 786.8 | 2.1 | 4 | 781.9 | 0.5 |
| 4(x1) | 778.2 | 3.1 | | | | | | |
| 5 | 747.2 | 0.7 | 5 | 745.4 | 2.0 | 5 | 745.5 | 1.0 |
| 6 | 704.7 | 1.3 | 6 | 707.5 | 1.1 | 6 | 706.6 | 1.1 |
| 7 | 665.6 | 2.8 | 7 | 664.7 | 1.0 | 7 | 661.7 | 1.1 |
| X | 640.3 | 2.3 | X | 641.5 | 1.6 | X | 637.9 | 1.4 |
| 8 | 605.3 | 1.5 | 8 | 604.3 | 1.4 | 8 | 604.8 | 1.6 |
| 9-12 | 560 | | 9-12 | 560 | | 9-12 | 560 | |
| 13 | 456.5 | 1.9 | 13 | 454.9 | 0.7 | 13 | 460.3 | 1.4 |
| 14 | 426.2 | 1.0 | 14 | 427.3 | 1.7 | 14, 15(x1) | 426.7 | 1.0 |
| | | | 15(x1)M | 413.6 | 2.3 | 15(x1)C | 415.7 | 0.6 |
| 15 | 406.0 | 1.2 | | | | | | |
| 16 | 388.2 | 1.4 | 16, 15del | 388.5 | 1.1 | 16 | 390.6 | 1.2 |
| 17 | 353.7 | 0.6 | 17 | 350.8 | 1.2 | 17 | 353.6 | 0.6 |
| 18 | 337.9 | 1.6 | 18 | 337.2 | 0.7 | 18(x1) | 341.5 | 1.8 |
| | | | | | | 18(x1) | 331.1 | 0.8 |
| 20 | 285.3 | 0.8 | 20,19(x1)M | 281.9 | 1.7 | 20,19(x1)C | 281.2 | 0.6 |
| 19 | 262.4 | 0.4 | 19(x1)P | 264.6 | 1.6 | 19(x1) | 263.2 | 0.9 |
| 22(x1)C | 232.1 | 1.3 | 22(x1)P | 230.9 | 1.2 | | | |
| 22(x1) | 251.1 | 0.9 | 22(x1)M | 217.7 | 0.9 | 22, 21(x1) | 217.9 | 1.2 |
| Y, 21(x1)C | 199.6 | 0.8 | Y, 21 | 196.7 | 0.5 | 21(x1)C | 192.5 | 0.8 |
| 21(x1) | 180.3 | 0.8 | | | | | | |

While the examination of flow karyotypes from all the members of a family will clearly be required in the majority of cases to aid the elucidation of potential chromosome abnormalities, variations in the observed pattern of an individual's flow karyotype for chromosomes which normally exhibit relatively little shift in relative fluorescence could be of some significance. An example of this in our department was provided by the patient whose flow karyotype is shown in Figure 5. Following initial cytogenetic examination, no chromosome abnormality was detected, but a large heteromorphism of chromosome 1 was observed. The patient's chromosomes were run on the flow cytometer with a view to sorting chromosome 1, and a flow karyotype was produced which suggested that one of the X chromosomes was smaller than normal (by approximately 5%). Subsequent cytogenetic re- examination revealed a deletion involving band q2.4.

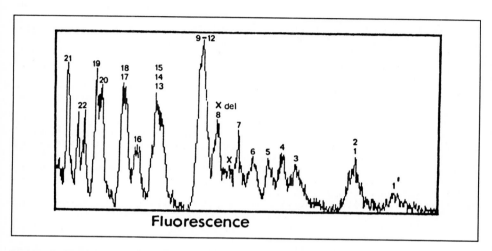

**Fig. 5.** Ethidium bromide flow karyotype of a female patient showing a large heteromorphic chromosome 1 and an X deletion.

## 4. Conclusion

Flow karyotype analysis can provide a useful adjunct to conventional cytogenetics in both the confirmation and quantification of certain chromosome abnormalities. The inability of flow analysis to distinguish between potential abnormalities and heteromorphism can be resolved, at least in part, by a detailed comparison of the flow karyotype of a child with those of his or her parents. The amount of time and resources required to produce high quality flow karyotypes would at present probably restrict this method of analysis to cases in which some form of detectable abnormality was clearly indicated either clinically or cytogenetically, rather than generalised screening of the typical sample intake of a routine medical genetics department.

*Acknowledgement.*The pictures of chromosome 15 shown in Figure 4 were provided by Jim Colgan.

# References

1   Collard JG, Phillippus E, Tulp A, Lebo RV, Gray JW: Separation and analysis of human chromosomes by combined velocity sedimentation and flow cytometry. Cytometry 5 9-19 (1984).

2   Cooke A, Tolmie J, Darlington W, Boyd E, Thomson RB, Ferguson- Smith MA: Confirmation of a suspected 16q deletion in a dysmorphic child by flow karyotype analysis. J Med Genet 24 88-92 (1987).

3   Van den Engh GJ, Trask BJ, Gray JW, Langlois RG, Yu LC: Preparation and bivariate analysis of suspensions of human chromosomes. Cytometry 6 92-100 (1985).

4   Gray JW, Carrano AV, Moore DH, Steinmetz LL, Minkler J, Mayall B, Mendelsohn ML, van Dilla MA: High speed quantitative karyotyping by flow microfluorometry. Clin Chem 21 1258-1262 (1975).

5   Green DK, Fantes JA, Buckton KE, Elder JK, Malloy P, Carothers A, Evans HJ: Karyotyping and identification of human chromosome polymorphisms by single fluorochrome flow cytometry. Hum Genet 66 143-146 (1984).

6   Harris P: Evaluation of the Fluorescence Activated Cell Sorter (FACS) for human karyotype analysis and chromosome sorting. PhD Thesis. University of Glasgow (1984).

7   Harris P, Boyd E, Young BD, Ferguson-Smith MA: Determination of the DNA content of human chromosomes by flow cytometry. Cytogenet Cell Genet 41 14-21 (1986).

8   Lalande M, Kunkel LM, Flint A, Latt SA: Development and use of metaphase chromosome methodology to obtain recombinant phage libraries enriched for parts of the human X chromosome. Cytometry 5 101-107 (1984).

9   Lebo RV, Golbus MS, Cheung MC: Detecting abnormal human chromosome constitutions by dual laser flow cytogenetics. Am J Med Genet 25 519-529 (1986).

10  Sillar R, Young BD: A new method for the preparation of metaphase chromosomes for flow analysis. J Histochem Cytochem 29 74-78 (1981).

11  Young BD, Ferguson-Smith MA, Sillar R, Boyd E: High resolution analysis of human peripheral lymphocyte chromosomes by flow cytometry. Proc Natl Acad Sci 78 7727-7731 (1981).

# Chromosome Aberration Detection with Hybridized DNA Probes: Digital Image Analysis and Slit Scan Flow Cytometry

Christoph Cremer, Michael Hausmann, Eduardo Diaz, Jutta Hetzel, Jacob A. Aten, Thomas Cremer

## Summary

Metaphase chromosomes of Chinese hamster × human hybrid cell lines were hybridized with biotin-labelled human total genomic DNA. The hybridized DNA and hence the human chromosomal material was visualized using either an alkaline phosphatase enzyme reaction or a FITC labelling detection system. For quantitative evaluation, both digital image analysis and slit scan flow cytometry were applied. The results suggest that the automatic detection of chromosome translocations with hybridized DNA probes is feasible both on slides and in suspension.

## 1. Introduction

An important field of automated cytogenetics is the detection of structural chromosome aberrations. While considerable progress has been made concerning the automated evaluation of dicentric chromosomes from homogeneously stained specimen (Gray and Langlois, 1986; Lörch and Stephan, 1986), the automated detection of other structural aberrations, such as translocations, is still in the beginning (Piper and Lundsteen, 1987; Willborn et al., 1987).

Here, an alternative to the "conventional" method based on the analysis of banded chromosomes is presented. The rationale of this new approach is the use of hybridization techniques to visualize specific chromosomes and chromosome subregions (Durnam et al., 1985; Manuelidis et al., 1985; Schardin et al., 1985; Cremer et al., 1986, 1988a; Pinkel et al., 1986; Emmerich et al., 1989). To exemplify the method, chromosomes from Chinese hamster × human hybrid cell lines are used. It is anticipated, however, that recent progress made in the visualization of specific chromosomes in human cells by in situ suppression hybridization using recombinant DNA libraries from sorted human chromosomes (Cremer et al., 1988b; Lichter et al., 1988) will make it feasible to use the approach, described here, also for the automated detection of chromosome rearrangements, such as translocations, in human cells.

**Automation of Cytogenetics**   Editors: C. Lundsteen  J. Piper
© Springer-Verlag Berlin Heidelberg New York 1989

## 2. Materials and Methods

For the evaluation of metaphase spreads, the Chinese hamster × human hybrid cell line GB3 (kindly provided by C.H.C.M. Buys, Groningen), containing 18 hamster chromosomes and two No. 13 chromosome equivalents as the only human material, was grown under standard conditions. Mitotic cells were enriched with demecolcine (Colcemid; $0.25\mu g/ml$ medium) and fixed on slides with methanol/acetic acid (3:1). The metaphase spreads were hybridized with biotinylated human total genomic DNA and the hybridized DNA was visualized with the alkaline phosphatase (AP) method essentially as described by Schardin et al. (1985) and Cremer et al. (1986).

For the analysis of isolated chromosomes in suspension, the Chinese hamster × human hybrid cell line $A_1wbf_2$ (kindly provided by P.Pearson, Leiden) was used, containing four human chromosomes and several interspecies translocations. The chromosomes were hybridized in suspension with biotinylated human total genomic DNA as described (Dudin et al., 1987a). Human material was detected due to the binding of FITC labelled antibodies (goat anti rabbit - rabbit anti biotin) or a streptavidin - FITC complex to the biotinylated hybridized DNA (Dudin et al., 1988; Hausmann et al., 1989).

In both cases the chromosomes were counterstained with propidium iodide (PI). Counterstaining with DAPI is also possible (Dudin et al., 1987b, Hausmann et al., 1988).

Digital image analysis of colour microphotographs of chromosomes on slides was performed using a Joyce-Loebl drum scanning densitometer (Scandig 2605) and a VAX 11/780 as described (Dudin et al., 1988, Hausmann et al., 1989).

Two parameter slit scan flow cytometry of isolated chromosomes in suspension was done on the Amsterdam slit scan flow cytometer (Aten et al., 1988, Hausmann, 1988). Briefly, the chromosomes pass a focussed laser beam (ca. $3\mu m$ diameter). The time dependent fluorescence signal is registered and thus produces a profile of the fluorescence distribution along the chromosome (Gray et al., 1979; Lucas et al., 1983; Lucas and Gray, 1987).

The system used is based on an Ortho Cytofluorograph. Simultaneous PI and FITC excitation were performed with an argon laser at 488nm with 500mW. The PI fluorescence was collected on channel 1 while the FITC emission was collected on channel 2 via a 525nm band pass filter.

## 3. Results

Figure 1 shows a metaphase spread of the Chinese hamster × human hybrid cell line GB3 following hybridization, alkaline phosphatase (AP) visualization of hybridized chromosome sites, and PI counterstaining. In this cell an interspecies translocation including a part of a human (black AP staining) and a Chinese hamster chromosome (PI staining) is clearly visible.

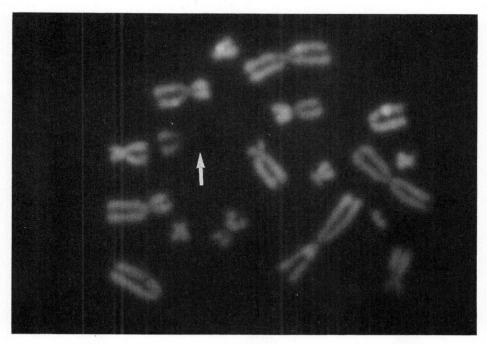

**Fig. 1.** Metaphase spread of a cell of the Chinese hamster × human hybrid cell line GB3 following in situ hybridization with biotinylated human total genomic DNA. The hybridized site indicating human material (arrow) is visualized by an alkaline phosphatase enzyme reaction (black staining).

Figure 2 shows the integrated density profiles of the chromosomes of the metaphase spread of figure 1. All Chinese hamster chromosomes have either a "bimodal" (e.g. figure 2a-e) or a "smooth" (e.g. figure 2 m-n) distribution as expected for homogeneous staining. The only "aberrant" profile observed is that of the interspecies translocation chromosome (figure 2j). In this case, two maxima of considerably uneven heights were observed. Thus, the translocation chromosome may be distinguished by quantitative characteristics. For example, the contrast

$$c = \frac{|\text{maximum 1} - \text{maximum 2}|}{\text{maximum 1} + \text{maximum 2}}$$

calculated for bimodal profiles of normal chromosomes was between 0.03 and 0.16 while for the profile of the translocation chromosome (figure 2j) a c-value of 0.36 was obtained. In case of a very strong AP reaction, PI fluorescence on the human part of the translocation chromosome may be suppressed completely. This may result in a "smooth" profile of the fluorescence detectable along the translocation chromosome. Additional information to distinguish such a case from chinese hamster chromosomes with a "smooth" fluorescent profile can

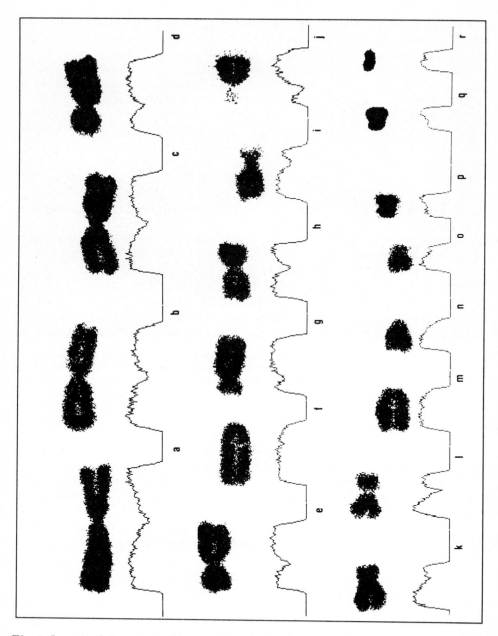

**Fig. 2.** Integrated density profiles (sum of gray level values perpendicular to the chromosome axis) of the chromosomes of figure 1. Chromosomes were numbered (a - r) according to length. For all chromosomes with "bimodal" profiles (a - i and k - l) the maximum of the profile on one arm is less than 1.4 higher than on the other. On profile j, however, the intensity on the long arm (Chinese hamster material) is 2.3 times higher than on the short arm (alkaline phosphatase labelled human part of the chromosome).

easily be obtained by digital image analysis of microphotographs taken with transmittant light which allows the direct measurement of accumulating dye due to the AP reaction.

In many cases translocation chromosomes are rare events. Therefore it may be important to accelerate the classification of chromosomes according to "normal" and significant "aberrant" profiles. Slit scan flow cytometry allows to acquire profiles of stain distribution along chromosomes (Lucas et al., 1987) at a rate up to 100/sec. Classification of the chromosomes is possible for instance according to the centromeric index (CI) of the fluorescence intensity profile. Therefore, it appeared to be interesting to examine whether this "one dimensional imaging" technique may be combined with the hybridization approach using a fluorochrome (FITC) detection system instead of AP reaction.

The morphology of chromosomes following fluorescence hybridization in suspension was well preserved (see also Dudin et al., 1987a). Integrated density profiles obtained for randomly selected FITC and/or PI stained chromosomes in suspension from several Chinese hamster × human hybrid cell lines were very similar to profiles obtained from conventional metaphase spreads (see also Hausmann et al., 1989). The rather homogeneous FITC staining pattern indicates a correspondingly homogeneous hybridization of human DNA sequences along the arms of isolated human chromosomes. Analogous results were obtained for the PI distributions along the Chinese hamster chromosomes of these cell lines.

For quantitative evaluation of CIs from integrated density profiles, different ways of calculation are possible (figure 3):

$$\text{(a) } CI(L) = \frac{\text{length of long arm}}{\text{entire chromosome length}},$$

$$\text{(b) } CI(A) = \frac{\text{area of long arm profile}}{\text{entire profile area}}$$

After staining with a DNA specific fluorochrome the entire profile area should be directly proportional to DNA content (and to chromosome length). Accordingly, there should not be any great variation in the linearity between $CI(L)$ and $CI(A)$.

Using these two modes of evaluation, CIs were obtained from chromosomes of the $A_1wbf_2$ cell line. As expected, in the case of a homogeneous binding of PI to the DNA of Chinese hamster chromosomes a linear relationship exists between $CI(A)$ and $CI(L)$ (figure 4b). The same linear relationship is also expected in the case of a homogeneous binding of the biotinylated human DNA to human chromosomes detected with FITC-streptavidin or a FITC labelled double antibody system. Figure 4a shows that this is indeed the case with high accuracy.

It is concluded that under the conditions used, a homogeneous FITC distribution on both chromosome arms corresponding to $CI(A)$ being proportional to $CI(L)$ will denote a normal chromosome. Alternatively, a non-homogeneous distribution of hybridized biotinylated human DNA, i.e. a strong

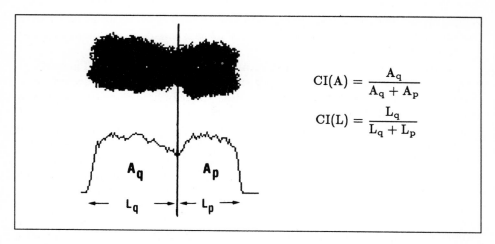

$$CI(A) = \frac{A_q}{A_q + A_p}$$

$$CI(L) = \frac{L_q}{L_q + L_p}$$

**Fig. 3.** Determination of the centromeric index (CI) from integrated density profiles (see text).

non-homogeneous FITC distribution, along the arms of a given target chromosome suggests a chromosome consisting of both Chinese hamster and human DNA (interspecies translocation).

Using two parameter slit scan flow cytometry, profiles of chromosomes ($A_1$wbf$_2$ cell line) were measured following fluorescence hybridization in suspension, FITC labelling, and counterstaining with PI (Hausmann et al., 1989). For every chromosome, FITC and PI fluorescence was measured simultaneously. In this case, the PI fluorescence distribution was used only to assure that the chromosome alignment was correct (i.e. ideally, in case of metacentric and submetacentric chromosomes a bimodal distribution is expected).

Figure 5 shows the PI-FITC dual parameter slit scan profiles of a presumably normal (a) and a presumably aberrant chromosome (b). The bimodal PI distribution indicates that the flow alignment was correct. In case (a), the symmetrical FITC distribution with two peaks indicates a profile of a "normal" chromosome; in case (b), however, the FITC profile is significantly asymmetrical. Only one FITC peak is clearly distinguishable in channel 2 and corresponds to the left PI peak in channel 1. A careful examination (Hausmann et al., 1989) of the slit scan profiles of about a hundred of correctly aligned chromosomes in combination with a microscopic evaluation led to the conclusion that aberrant profiles of this type are indeed due to structural chromosome aberrations (in this case an interspecies translocation).

## 4. Discussion

Recently, it has been shown (Durnam et al., 1985; Schardin et al., 1985; Pinkel et al., 1986) that interspecies translocations in metaphase spreads of Chinese

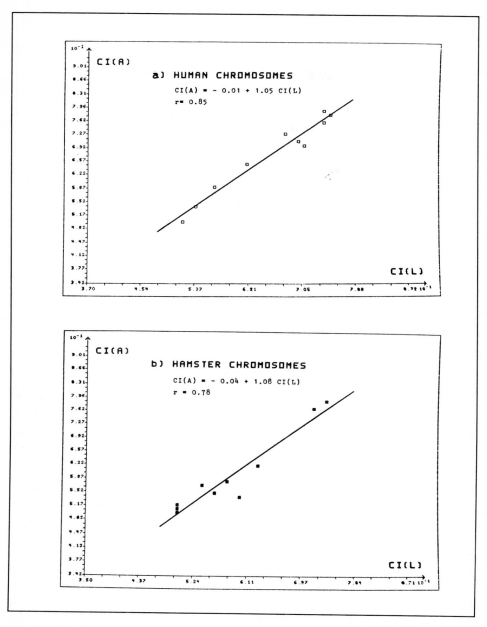

**Fig. 4.** Ordinate: CI(A): Centromeric index as determined from density profile areas (see figure 3 and text) Abscissa: CI(L): Centromeric index as determined from profile length (see figure 3 and text) a) Relationship between CI(A) and CI(L) for FITC fluorescing chromosomes in suspension (human chromosomes) b) Relationship between CI(A) and CI(L) for PI fluorescing chromosomes in suspension (chinese hamster chromosomes)

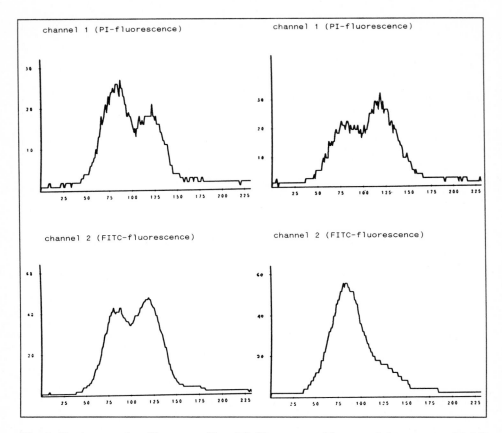

**Fig. 5.** Dual parameter slit scan profiles of (left) a presumably normal chromosome, (right) a presumably aberrant chromosome. Ordinate: Fluorescence intensity (channel 1: PI fluorescence, channel 2: FITC fluorescence). Abscissa: Time of flight (in units of 100nsec)

hamster × human hybrid cells may easily be assessed by microscopic evaluation following in situ hybridization with biotinylated human genomic DNA. No image analysis, however, was performed in these studies.

The results presented here suggest that automated detection of translocations following hybridization with specific DNA probes is feasible. This is shown both by conventional digital image analysis of microphotographs of metaphase spreads hybridized on slides and by two parameter slit scan flow cytometry following fluorescence hybridization of isolated chromosomes in suspension. So far, the method has been used for interspecies translocations. These are by themselves interesting, e.g. for dosimetry of radiation induced aberrations (Pinkel et al., 1986). Recently, however, chromosome specific DNA libraries (Van Dilla et al., 1986) were used to visualize specifically individual metaphase chromosomes in human normal and aberrant cells by in situ suppression hybridization techniques (Cremer et al., 1988b; Lichter et al., 1988).

This will allow to extend the approach described here to the automated detection of structural chromosome aberrations in human cells, e.g. for monitoring of radiation damage, or application in constitutional and tumor cytogenetics.

*Acknowledgements.* This study was supported by the Deutsche Forschungsgemeinschaft. T. Cremer is the recipient of a Heisenberg stipendium. We thank the Institute of Nuclear Medicine (German Cancer Research Center, Heidelberg) for the possibility to use the Joyce Loebl drum scanning densitometer and the VAX 11/780 computer for digital image analysis. We also thank P. Zuse for some help in software.

# References

1.   Aten JA, Manders E, van Owen C, Reus W, Stap J. Detection of chromosome aberrations by flow cytometry as a test of radiation sensitivity. Cytometry, Suppl. 2:23 (1988)
2.   Cremer T, Landegent J, Brückner A, Scholl HP, Schardin M, Hager HD, Devilee P, Pearson P, van der Ploeg M. Detection of chromosome aberrations in the human interphase nucleus by visualization of specific target DNAs with radioactive and non-radioactive in situ hybridization techniques: diagnosis of trisomy 18 with probe L1.84. Hum. Genet. 74:346-352 (1986)
3.   Cremer T, Tesin D, Hopman AHN, Manuelidis L. Rapid interphase and metaphase assessment of specific chromosomal changes in neuroectodermal tumor cells by in situ hybridization with chemically modified DNA probes. Exp. Cell Res. 176:199-220 (1988a)
4.   Cremer T, Lichter P, Borden J, Ward DC, Manuelidis L Detection of chromosome aberrations in metaphase and interphase tumor cells by suppression hybridization using chromosome specific library probes. Hum. Genet. 80:235-246 (1988b)
5.   Van Dilla MA et al. Human chromosome - specific DNA libraries construction and availability. Biotechnology 4:537-552 (1986)
6.   Durnam DM, Gelinas RE, Myerson D. Detection of species specific chromosomes in somatic cell hybrids. Somatic Cell Mol. Genet. 11:571-577 (1985)
7.   Dudin G, Cremer T, Schardin M, Hausmann M, Bier F, Cremer C. A method for nucleic acid hybridization to isolated chromosomes in suspension. Hum.Genet. 76:290-292 (1987a)
8.   Dudin G, Hausmann M, Rens W, Aten JA, Cremer C. Fluorescence hybridization of isolated metaphase chromosomes in suspension for flow cytometric analysis. Ann. Univ. Sarav. Med., Suppl. 7:81-84 (1987b)
9.   Dudin G, Steegmayer EW, Vogt P, Schnitzer H, Diaz E, Howell K, Cremer T, Cremer C. Sorting of chromosomes by magnetic separation. Hum. Genet. 80:111-116 (1988)
10.  Emmerich P, Loos P, Jauch A, Hopman AHN, Wiegant J, Higgins M, White BN, van der Ploeg M, Cremer C, Cremer T. Double in situ hybridization in combination with digital image analysis: A new approach to study interface chromosome topography. Exp. Cell Res. (in press, 1989)
11.  Gray JW, Langlois RG. Chromosome classification and purification using flow cytometry and sorting. Ann. Rev. Biophys. Biophys. Chem. 15:195-235 (1986)
12.  Gray JW, Peters D, Merrill JT, Martin R, van Dilla MA . Slit scan flow cytometry of mammalian chromosomes. J.Histochem.Cytochem. 27:441-444 (1979)
13.  Hausmann M. Laserfluoreszenzaktivierte Analyse und Sortierung von Metaphasechromosomen, Anwendung der Slit-Scan Flussphotometrie, Dissertation, Fakultät für Physik und Astronomie, University of Heidelberg (1988)

14. Hausmann M, Dudin G, Aten JA, Bier F, Cremer C. Slit - scan flow cytometry following fluorescence hybridization: A new approach to detect chromosome translocations. Eur. J. Cell Biol. 46 (Suppl. 22):26 (1988)

15. Hausmann M, Dudin G, Aten JA, Bühring H-J, Diaz E, Dölle J, Bier FF, Cremer C. Flow cytometric detection of isolated chromosomes following fluorescence hybridization. Biomedical Optics (in press, 1989)

16. Lichter P, Cremer T, Borden J, Manuelidis L, Ward DC. Delineation of individual human chromosomes in metaphase and interphase cells by in situ suppression hybridization using recombinant DNA libraries. Hum.Genet. 80:224-234 (1988)

17. Lörch T, Stephan G. Aufbau eines Systems zur automatischen Erkennung dizentrischer Chromosomen. ISH - Heft 89 (Institut für Strahlenhygiene des BGA, D-8042 Neuherberg) (1986)

18. Lucas JN, Gray JW. Centromeric index versus DNA content flow karyotypes of human chromosomes measured by means of slit scan flow cytometry. Cytometry 8:273-279 (1987)

19. Lucas JN, Gray JW, Peters D, van Dilla MA. Centromeric index measurement by slit scan flow cytometry. Cytometry 4:1O9-116 (1983)

20. Manuelidis L. Individual interphase chromosome domains revealed by in situ hybridization. Hum. Genet. 71:288-293 (1985)

21. Pinkel D, Straume T, Gray JW. Cytogenetic analysis using quantitative, high-sensitivity, fluorescence hybridization. Proc. Natl. Acad. Sci. (USA) 83:2934-2938 (1986)

22. Piper J, Lundsteen C. Human chromosome analysis by machine. Trends in Genetics 3:309-313 (1987)

23. Schardin M, Cremer T, Hager HD, Lang M. Specific staining of human chromosomes in Chinese hamster x human cell lines demonstrates interphase territories. Hum.Genet. 71:281-287 (1985)

24. Willborn K, Cremer T, Hausmann M, Cremer C. An approach to analysis of structural aberrations on the basis of schematic representations of chromosomes by a personal computer. Eur. J. Cell Biol. 43, Suppl. 17:64 (1987)

Part IV

**Automatic Preparation
of Cytogenetic Specimens**

# An Automated System for the Culturing and Harvesting of Human Chromosome Specimens

J. Vrolijk, G. Korthof, G. Vletter, C.R.G. van der Geest,
G.W. Gerrese, P.L. Pearson

## Summary

This paper describes a system for the automated culturing and harvesting of human chromosome specimens. The machine is capable of handling different preparation methods simultaneously, such as standard blood cell, blast cell and bone marrow cultures, and can be programmed to perform the prophase synchronization techniques. It is composed of a culture tray, centrifuge, mixer, input and output station and a head assembly capable of transporting samples between the various stations. The head is equipped with an aspirator needle to aspirate the supernatant and fluid dispensers for the dosing of the various chemicals. A microprocessor system controls all hardware functions and schedules the manipulations of all samples. The samples (up to a maximum of 255) are processed by the machine in small batches of at most 16, which can be activated at any time according to one of 16 culture procedures. The cytogeneticist can modify these procedures using a simple interpretive language specifying both the types of manipulations, such as centrifugation, addition of chemicals, aspiration of supernatant, and the minimal and maximal variation in time which is allowed between two sequential manipulations. Besides the hardware setup and software organization of the machine, the first preliminary results and future prospects for the machine are presented.

## 1. Introduction

In the past two decades considerable effort has been devoted to the automation of chromosome analysis. Most of the work was thereby dedicated to metaphase finding, both interactive and automated karyotyping and the automated detection of chromosome aberrations caused by environmental insults. Relatively little attention has been paid to the automation of culturing, harvesting and slide preparation. Although the latter aspects seem to have been neglected, these procedures involve relatively simple manipulations, which can be easily mechanized and do not require any human evaluation. Another argument in favour of automation of preparation techniques is the rather monotonous work pattern involved, while it still takes approximately forty percent of a technician's time.

**Automation of Cytogenetics**   Editors: C. Lundsteen  J. Piper
© Springer-Verlag Berlin Heidelberg New York 1989

So far a few systems have been described in literature concerning aspects of chromosome preparation. The Tecan system [1] and the one described by Wulf [2] are simple semi-automated machines to increase the efficiency of harvesting procedures. The City of Hope (CoH) system developed by Melnyk et al. [3] and tested by Shaunnessey et al. [4,5] and the Chromosomat developed by Stranzinger [6] are automated systems. The CoH system handled the aspects of culturing, harvesting and slide preparation, while only the harvesting step is performed by the Chromosomat. These last two systems can handle a large number of samples simultaneously but only according to one specific procedure at a time.

In our opinion such a harvesting machine should cope with a more realistic image of the situation within a cytogenetics laboratory. In the daily routine samples of different cell types from different sorts of patients come in by post or directly from the hospital in small quantities and have to be treated differently. Thus a machine should rather allow the simultaneous processing of small sets of samples according to various culturing methods. In our laboratory blood, lymphoblasts and bone marrow cultures are carried out routinely, but also the prophase synchronization techniques, which require the addition of extra chemicals (methotrexate and thymidine or BrdU), the combined acridine orange-colcemid incubation, and FUdR incubation for fragile site detection. Due to the variety of culture methods the machine should allow flexible and easy adaption of these procedures.

As the best quality of slides is obtained by performing the preparation method on fresh material, the machine should not impose restrictions on the time when samples are put into the machine. Thus, besides the simultaneous handling of different methods, the machine should also allow the asynchronous activation and control of the culture procedures. Especially this latter feature illustrates the advantage of automation of culturing and harvesting. As there are optimal culture times (for a standard blood culture, ca. 72 h) and as there are normally no weekend services in our laboratory, not all incoming samples can be optimally processed manually. However, a machine without such time restrictions can continue operating during the weekend or at night.

As the machine is basically developed for routine use, the capacity should be large enough to handle all the material of the patients. This means, that for our laboratory, which with 1200 patients per year can be considered as being a medium sized cytogenetics laboratory, the machine should perform about 6000 cultures per year, since on the average 5 cultures per patient are carried out.

The requirements which the culture machine should fulfil to cope with the situation in our laboratory have strongly influenced both the mechanical design, structure and organization of the system hardware and software.

# 2. Materials and methods

## 2.1. Hardware aspects

Figure 1 shows a schematic diagram of the culture procedures, which should be processed by the system. This diagram suggests that only a few basic manipulations need to be carried out by such a machine, namely: culturing at a controlled temperature, centrifugation with adjustable speeds, the addition of chemicals in different volumes and with different flow rates, aspirating the supernatant, and resuspension of the pellet.

The system was constructed in such a way that it can only handle one sample at a time. Basically it consists of a head capable of gripping and transporting a test tube to the various stations. The head is equipped with dispensers for the adding of chemicals on a mixer and a needle for aspirating supernatant. The processing of a batch of samples is carried out by the machine in a similar way compared to the manual routine, where the technician adds the hypotonic fluid or the fixative to the samples one after each other and then centrifuges the set simultaneously, so that the time between adding chemicals and centrifugation is different for each sample. As the machine is slower compared to the technician (typical time for transport between centrifuge and mixer, aspirating the supernatant and adding chemicals during mixing is 20sec), these differences might result in slides of poor quality when the batch size increases. As in practice, however, only small sets of samples have to be handled simultaneously, this is not a severe limitation. Accordingly the system has been designed to process the samples in batches of at most 16 test tubes. As an illustration the overall processing time for a harvesting procedure of lymphocyte blood cultures is shown as a function of the batch size in figure 2.

The system consists of the following parts:

1. The culture tray. The culture tray is a square aluminum plate with a heating foil coupled to the bottom. This foil provides an equable temperature over the plate. The temperature is controlled by the computer at 37.5°C. The tray has 256 holes, in which samples can be cultured. One hole is reserved for a dummy test tube, which is used to balance the centrifuge in case of an odd number of samples.
2. The centrifuge. The centrifuge is equipped with 4 buckets, each of which can contain 4 samples. Thus a batch of up to 16 samples can be processed by the system. The maximum speed of the centrifuge is 1500 rotations per minute. It is controlled by a servo-motor provided with an optical encoder, which can position the buckets with an accuracy of 0.18°.
3. The mixer. A normal commercially available Vortex mixer has been applied to resuspend the pellet of the samples. Only the on/off switch is controlled by the computer.

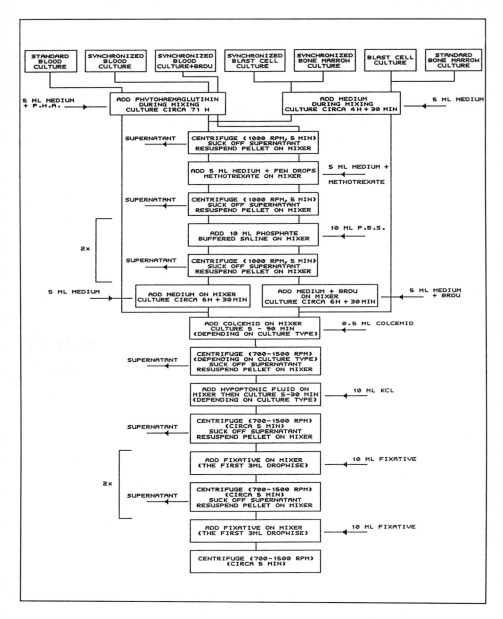

**Fig. 1.** Schematic diagram of the various culture procedures carried out in our laboratory.

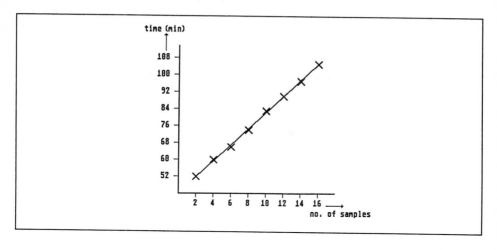

**Fig. 2.** The processing time of a harvesting procedure as function of the batch size.

4. The input and output station. The sample input unit is provided with 16 holes, so that batches of up to 16 samples can be activated. The output unit is provided with 40 holes, in which the samples are placed after harvesting. As samples might be completed at night or during the weekend and as the quality reduces when these samples are kept at room temperature, the output unit is constantly cooled at a temperature of 4°C.

5. The head assembly. The head consists basically of two units. The first one lifts and lowers the gripper of the caps on the test tubes, which protect the cultures against contamination. The caps are gripped by sucking vacuum. Before each manipulation the cap is removed and afterwards replaced on each tube. The second unit also pneumatically controls the lifting, lowering, gripping and releasing of the tube itself. The gripper consists of a ring with 3 equally spaced balls, which grips the tubes sufficiently firmly and flexibly to permit mixing on a Vortex mixer. The head is furthermore equipped with a needle, which can be lowered into the tube to aspirate the supernatant. Around this aspirator needle 7 tubes pass down through the head and are used for adding various liquids to the cultures. A schematical diagram of the head is shown in figure 3.

6. The transport mechanism. The head unit is moved to the various stations by a transport mechanism consisting of 2 screw-spindles, one in the X and one in the Y-direction, which are driven by servo-motors equipped with optical encoders. The head can be positioned with an accurancy of 0.1 mm.

7. Fluid dispensers. The addition of liquids is carried out using so called metering valves. This metering valve can be opened by a timed air pulse from, at the minimum, 20msec. By changing the pulse length the flow

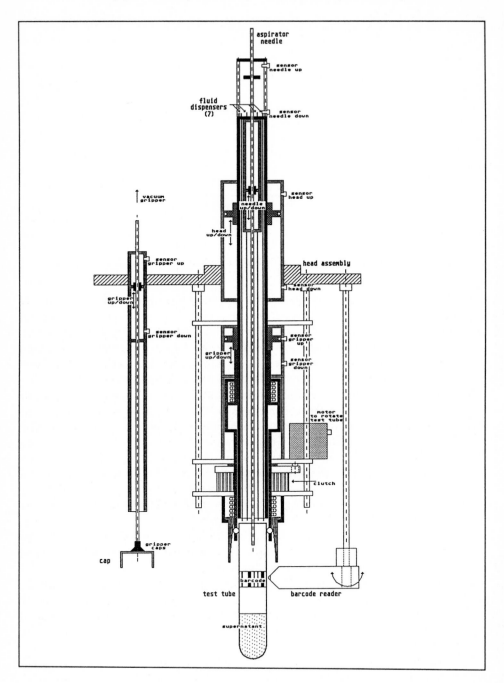

**Fig. 3.** Schematic diagram of the head.

rate per second can be varied. The actual range of the flow rate is also dependent on the air pressure in the supply vessel and the viscosity of the liquid. It has been adjusted to flow rates ranging from $50\mu l$ to 2ml per second. The supply vessels are provided with volume sensors, so that the user can see how much liquid is still present, and the system can check if the required addition can be carried out. Before each fixation the total amount of fixative is made for all the samples of the batch by mixing methanol and acetic acid in the right proportions.

8. The waste vessel. The supernatant is transported to a small waste-vessel placed on the head assembly and later on automatically to a large waste-vessel down in the machine, which has to be emptied only once a week.

9. The controller. A microprocessor system based on a 6809 processor controls all hardware functions of the machine and schedules the manipulations of the samples. A block diagram of the system is shown in figure 4.

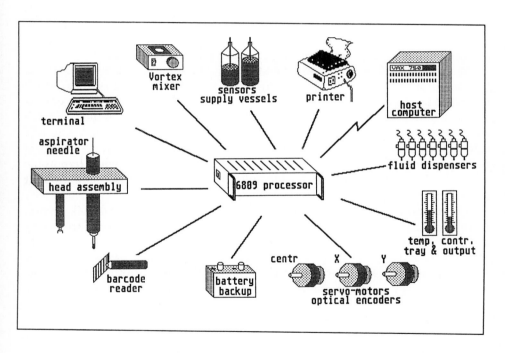

**Fig. 4.** Configuration of the system.

10. Battery backup. In order to increase the reliability of the system battery backup is provided for both the microprocessor system and the pneumatics. When a power failure occurs, processing is suspended only at the moment, that centrifugation or transportation of the samples is required, as the

servo-motors are not equipped with backup provisions due to the high costs involved. When the power comes up again, processing is automatically resumed. However, the time that a power failure lasts is logged in an error queue, so that the user can evaluate if it would have influenced the results.

A photograph of the current prototype is shown in figure 5.

## 2.2. Software aspects

A foreground/background approach was chosen for the software system, where time critical processes, such as control of the hardware functions and execution of instructions for the samples are handled by the foreground, whereas user commands, such as editing facilities are carried out in the background.

The foreground is activated using a real time clock, which generates interrupts at a frequency of 1000 Hz. On each interrupt, control is passed to a separate process, which is responsible for the functioning of a separate task in the machine. Examples of such processes are: the control of the servo-motors for the positioning of the head-assembly and the centrifuge, the sampling of the temperature controllers, the centrifugation of a batch of samples, the updating of the current date and time and the interpretation and execution of statements for the samples. The actual manipulations on the samples are also controlled by a foreground process. They can be subdivided into a series of actions, which are carried out after each other. For example, the picking up of a tube consists of the positioning of the head assembly above the tube, the removal of the cap, the lowering of the head and finally the gripping of the tube. Using this software approach a lot of tasks are carried out simultaneously, so that it is possible in practice that the pellet of a sample is being mixed, while hypotonic fluid is added and another batch of samples is centrifuged, while a new amount of fixative is being made.

When no foreground job is active, control is returned to the background. In the background the user can create, edit or delete culture procedures using a simple line editor. The system can currently hold 16 different procedures resident in memory. The samples can be activated in small batches to one of these procedures. At the moment of activation the technician can specify the values of the different parameters, such as time for hypotonic treatment, only for the batch concerned, so that there is still an amount of variation possible between batches which are carried out according to the same procedure. As no mass storage devices are present on the culturing machine itself, the procedures can be stored or loaded from the host computer, currently a VAX-11/750, using a RS-232 serial link.

The user specifies in a statement of a culture procedure everything that should happen between picking-up and releasing of the sample, and also defines which of the manipulations should be carried out in parallel and which after each other. The construction is similar to the "fork" and "pipe" constructs of the Unix operating system. For example:

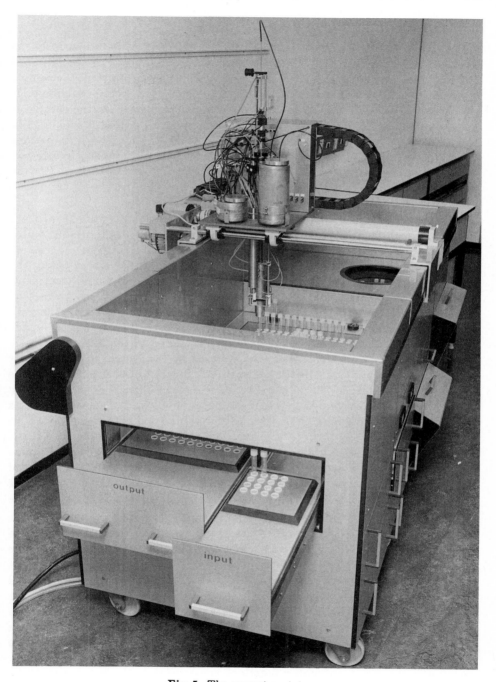

**Fig. 5.** The current prototype.

}MIX & ADD 0,1000,50; MIX ,10;}CEN

specifies that the sample should go to the mixer, while simultaneously the fluid controlled by valve 0 is injected with a speed of 50 drops per second, until 1000 $\mu$l have been added. When both manipulations are completed, the sample is mixed for 10 sec at the Vortex and finally transported to the centrifuge. It should be noted that the user does not have to specify where the sample is located at the beginning of the statement and to which position it should go in the centrifuge. The system automatically includes the necessary manipulations during execution and keeps track of the positions of all the samples. In this way a procedure is obtained which does not contain any information about specific positions, and is therefore applicable for each sample which is being processed according to that procedure.

Normally between two instructions which require the use of the head, there is at least one statement which does not use the head, as otherwise the user could have programmed the two instructions more efficiently in one "pipe" statement. Mostly it is necessary that a certain time should elapse before the next manipulation takes place. For instance, after the hypotonic treatment, centrifugation is carried out 20min later. The user can specify such times by means of the "DELAY" instruction, which defines the minimum time that will elapse before the next statement is carried out. However, the user has to include also the maximal variation in time, that can be added to the fixed delay. The variable parts of "DELAY" statements are used by the system to schedule not only the manipulations of the samples within a batch, but also to schedule the batch in accordance to the already active samples. This is done as follows: when a batch is activated, the system estimates how long each statement lasts and checks whether the batch of samples can be handled within the variations of time specified in the procedure. Then the batch is fitted into the time table, which holds information about when the head and the centrifuge are occupied by the already active samples. This is performed in such a way that the head is never required for more than one sample simultaneously, and the centrifuge never for more than one batch simultaneously. In figure 6 the listing of a simple harvesting procedure is shown.

The software system provides furthermore a set of utilities which allow the user to inspect the current state of the machine, to inform when samples will be ready, to look at temperature distribution of tray and output unit, and to list the error queue, in which each error is logged with the time when it occurred and the sample that it concerned. Finally a set of privileged commands are available for the system operator. These allow the testing of each separate part of the machine, such as positioning of the head assembly at any absolute position and control of the pneumatics to lower the aspirator needle, lift the head, etc.. These functions ease the repair when something is malfunctioning. When any of these maintenance functions is carried out, the processing of the manipulations on samples is postponed, so that the head assembly does not

```
 1.  DELAY „ ; VARIA ,16,

 2.  CEN 1023,5

 3.  DELAY „ ; VARIA ,16,

 4.  STATUS hypotonic treatment

 5.  }MIX & SUCK 75; MIX ,1; ADD 0,2000,6,5000,50; MIX ,1; }TRAY

 6.  DELAY ,20,; VARIA ,15,

 7.  CEN 1000,6

 8.  DELAY „ ; VARIA ,16,

 9.  STATUS fist fixation

10.  }MIX & suck 75; MIX ,1; ADD 1,1000,4,1000,6,8000,50; MIX,1; }CEN

11.  DELAY „ ; VARIA ,16,

12.  CEN 1023,5

13.  DELAY „ ; VARIA ,16,

14.  STATUS second fixation

15.  }MIX & SUCK 75; MIX ,1; ADD 1,1000,6,8000,50; }CEN

16.  DELAY „ ; VARIA ,16,

17.  CEN 1023,5

18.  DELAY „ ; VARIA ,16,

19.  STATUS third fixation

20.  }MIX & SUCK 75; MIX ,1; ADD 1,1000,6,8000,50; }CEN

21.  DELAY „ ; VARIA ,16,

22.  CEN 1023,5

23.  DELAY „ ; VARIA ,16,
```

**Fig. 6.** Listing of a harvesting procedure.

start moving when the screw-spindles are being oiled, or the system does not try to add hypotonic when the corresponding supply vessel is being refilled.

## 3. Results

During the development of the culturing machine several pilot studies were carried out to establish and optimize the functioning of the various parts of the machine, such as culturing in the tray, and methods for the addition of liquids. This has resulted in a standard procedure for the automated harvesting of lymphocyte blood cultures. A total of 144 cultures were initiated according to this procedure. In the first 5 months of the test period 51 out of 89 (57.3%) cultures were succesfully completed by the machine and in the second 5 months 51 out of 55 (92.7%). Since at the occurence of one error the complete processing is stopped because of safety precautions, the success rate would have been considerably higher if the samples had been processed in smaller batches or if human interaction had been applied.

In order to evaluate the quality of the cultures 65 slides were compared to the slides obtained by the manual control. An adapted version of the metaphase finder described by Vrolijk et al. [7] was applied using the Leytas system [8]. This program, called MIAREA, automatically selects the metaphases on a slide, measures the area of the mask of each metaphase as a measure of spreading and records the number of non-dividing nuclei, so that the mitotic index (the ratio between the number of metaphases and non-dividing nuclei) can be calulated in order evaluate eventual selective metaphase loss caused by the harvesting procedure. False positives, such as dirt, are interactivally eliminated.

The mean metaphase area and mean mitotic index for all specimens are shown in table 1. The quality of the metaphases appeared to be more or less the same for machine and manual control. Also the correlation coefficient between machine and manual control was calculated for the mean metaphase area (firstly averaged per slide) and was 0.61, which is statistically significant. (A slide of bad quality from the control corresponds with a slide of bad quality from the machine of the same patient and the same holds for slides of good quality). Table 1 shows also that the mitotic index of the manual control is somewhat higher compared to the machine results, although the standard deviation is rather high. Regarding mitotic index, no statistically significant correlation could be found between the results of machine and manual control.

**Table 1.** Results.

| applied technique | mean area metaphases | standard deviation | mitotic index | standard deviation |
|---|---|---|---|---|
| machine | 447.3 | 95.5 | 1.38 | 0.65 |
| manual | 428.3 | 92.3 | 1.87 | 1.00 |

Although most of the effort so far has been spent to develop and optimize a procedure for lymphocyte blood, some samples were also succesfully processed using the FUdR/BrdU synchronisation techniques, and even a mouse cell line was harvested by the machine. However, the number of samples is too low to draw any conclusion from this experiment.

# 4. Discussion

A machine for the culturing and harvesting of chromosome specimen has been developed, which is capable of handling small batches of samples according to 16 different culture procedures, while it does not impose time restrictions to initiate these procedures. Flexibility is thereby maintained, as parameters of the culture procedures, such as the time for hypotonic treatment, and the amount of fixative required, still can be adapted by the technician for every batch of samples. This closely relates to the clinical routine of a cytogenetics laboratory.

Up to now a reasonable number of lymphocyte blood cultures has been handled automatically. The results show that the same slide quality can be achieved with this machine compared to the manual control and that no signifant loss of metaphase spreads occurs.

Considering the reliability of the system, most of the errors were caused by the fact that normal machine operation interferes with new developments of hardware and software, and that due to the complexity of the system the occurences of some errors are often so rare, that it takes a long time to solve them. Mechanical failures were also encountered, such as unsmooth movement of the aspirator needle, the loosening of caps and test tubes, and position errors of the servo-motors. All errors known so far have been more or less satisfactory eliminated.

Regarding the safety aspects, it should be mentioned that besides the usual checks in software to prevent malfunctioning, hardwired circuits are incorporated to protect the system against self-destruction. It is for instance impossible to position the head assembly when the head is down, or to lower the head when the servo-motors are active. Whenever a system failure occurs, the system is brought into a safe state with all fluid dispensers closed, so that for instance chemicals will not make a mess of the machine during the night. It should also be noted, that since blood samples might be contaminated and as aerosols might be produced during centrifugation, the caps stay on the test-tubes during centrifugation. The waste (supernatant) is automatically collected in the waste-bucket and discarded in the same way as all other blood-products in our laboratory.

In the future the potentials of the machine will be exploited more fully by also harvesting bone marrow specimens, and by applying synchronisation techniques, which will encompass more than 8 hours a day. Only when such tests have proven to be successfull can the machine be incorporated into a clinical routine.

# References

1. Spureck J, Carlson R, Allen J, Dewald G. Culturing and robotic harvesting of bone marrow, lymph nodes, peripheral bloods, fibroblasts, solid tumors with in situ techniques. Cancer Genetics and Cytogenetics 32:59-66 (1988)
2. Wulf HC. Mechanical Preparation of Cells for Chromosome Studies. Hum. Hered. 25:398-401, Chromosomes Today, Vol 5 (1975)
3. Melnyk J, Persinger GW, Mount B, Castleman KR. A semi- automated specimen preparation system for cytogenetics. Proceedings of automation of cytogenetics, Asilomar workshop, 51-67 (1975)
4. Shaunnessey MS, Martin AO, Sabrin HW, Cimino MC, Rissman A. A new era for cytogenetics laboratories: automated specimen preparation. Proceedings of the fifth annual symposium on computer applications in medical care, pp.538-542 (1981)
5. Martin AO, Shaunnessey MS, Sabrin HW, Maremont S, Dyer A, Cimino MC, Rissman A, McKinney RD, Cohen MM, Jenkins E, Kowal D, Dunn JK, Simpson JL. Evaluation and Development of a System for Automated Preparation of Blood Specimens for Cytogenetic Analysis. This volume.
6. Stranzinger GF. General aspects on the automation of chromosome preparations. Proceedings of the 4th European chromosome analysis workshop, Edinburgh, pp.7.2.1-7.2.3 (1981).
7. Vrolijk J, Pearson PL (1980) Video techniques applied to chromosome analysis. Microsc. Acta, Suppl. 4:105-115 (1980)
8. Vrolijk J, Pearson PL, Ploem JS. Leytas, a system for the processing of microscopic images. Anal. Quant. Cytol. 2:41-48 (1980)

# Evaluation and Development of a System for Automated Preparation of Blood Specimens for Cytogenetic Analysis

Alice O. Martin, Michael Shaunnessey, Howard Sabrin,
Sheri Maremont, Alan Dyer, Michael C. Cimino, Amy Rissman,
R.D. McKinney, Maimon M. Cohen, Edmund C. Jenkins,
Dean Kowal, Joe Leigh Simpson

## Summary

The CASPS automated chromosome preparation system was evaluated and developed over a period of years, resulting in a machine and protocol for its use that were capable of handling a large throughput of specimens. The slides produced from blood specimens were of similar quality to manually prepared slides, and the machine could be operated by a cytogenetically inexperienced operator. Cost-benefit analysis showed that overall labor costs were reduced by 64% compared to manual preparation.

## 1. Introduction

There are two phases of cytogenetic laboratory procedures: specimen preparation and analysis. Automation in the analysis phase (location of cells in metaphase, counting chromosomes, and karyotyping) has become an integral part of many clinical cytogenetic laboratories (Lundsteen and Martin, 1988). However, comparable progress has not been made in incorporating automated specimen preparation (culture initiation and "harvesting"). A few systems semiautomate the harvesting steps.* The TECAN performs the fluid aspiration and dispensing steps and is used routinely in some laboratories (Spureck et. al., 1988). Computer control permits adjustment of various parameters e.g. time the cultures are in various fluids, speed of pipette movement. Petri dishes and chamber slides are the culture vessels which can be used for bloods, amniotic fluids, and bone marrows. Another system (described in this publication) performs culture intitiation and harvesting.

---

* TECAN, U.S. Ltd., Chapel Hill, U.S.A. and Hombrechtikon, Switzerland; GI-RO 3283 Multislide Machine, MA.re, Milan, Italy; CROMAT 90, S&H Milan, Italy; Etaleur Chaffant Type Ect 85, Productions John Toulemonde & Cie, Paris, France; Machine Lab Tech, Sermeter, Niles, IL, U.S.A.

**Automation of Cytogenetics**   Editors: C. Lundsteen J. Piper
© Springer-Verlag Berlin Heidelberg New York 1989

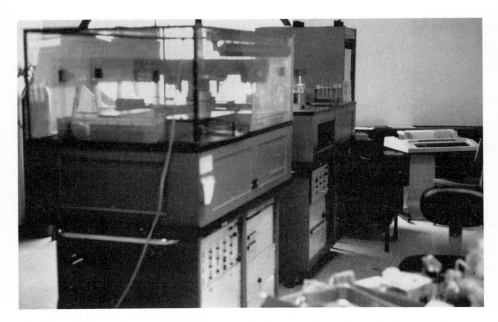

**Fig. 1.** The Clinical Automated Specimen Preparation System (CASPS). Console 1 is in the foreground; Console 2, the PDP 11/34 and the Decwriter printer are in the background. The monitor (CRT) is not shown.

Expanding indications and options for diagnostic cytogenetic analysis are dramatically increasing the number of patients referred for this service. Prenatal diagnosis is now available from 9 weeks of gestation by chorionic villus sampling. Because it is performed early in pregnancy and results are available within 1 week, this test is very popular. Low values of maternal serum alphafetoprotein are new indications for amniocentesis because of associated risk of trisomies. Fragile X testing is available for mentally retarded males and their families. Presence of a fragile X is a major cause of mental retardation. Other indications for cytogenetic analysis are cancer, in particular hematologic malignancies, and exposure to environmental agents which may damage chromosomes. Because of these increasing demands on laboratories, it is not suprising that automated metaphase locating and karyotyping systems are finding widespread acceptance. It seems timely also to consider further development of automated specimen preparation systems to further streamline laboratory procedures. For those considering such a venture, we present our experience in evaluating and developing a prototype system that prepared microscope slides from blood specimens for cytogenetic analysis.

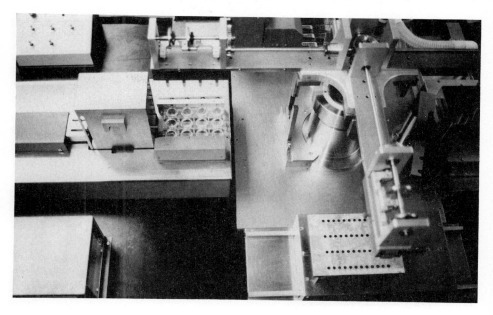

**Fig. 2.** Aerial view of one half of Console 1. Three of the 4 arms of the rotating carousel are visible. One tray of 12 culture wells are visible on the left at Station 2. In the foreground (Station 1) the holder for 48 pipette tips is shown. The Console is set for slide dropping. Bar code labelled slides are in the background at Station 3, a collector for used pipettes at Station 4. For culture initiation, 3ml vacutainers would be in Station 1.

## The Specimen Preparation System

The Clinical Automated Specimen Preparation System (CASPS), was built by Castleman and Melnyk, (1975) at the Jet Propulsion Laboratory and the City of Hope, California (Melnyk et al., 1970). One goal of developing this system was to prepare cleaner, more uniform slides to facilitate automated metaphase locating by their prototype automated light microscope, the "JPL system". Another stimulus was the expectation that widespread population screening for environmental damage to the chromosomes would come into vogue, necessitating the batch processing of large numbers of blood specimens. Controls and replicate cultures could be processed simultaneously.

The U.S. National Institutes of Health (NIH) put out requests for proposals to evaluate the automated systems and incorporate them into a clinical environment. Approximately 20 groups responded. After review of proposals, NIH transferred the CASPS to the Laboratory of Cytogenetics, Section of Human Genetics, Department of Obstetrics and Gynecology at Northwestern University, where it was evaluated in a clinical setting and further developed, 1978–81 (Martin, 1981, 1985; Martin and Unger, 1985, Martin et al, 1986, Sabrin et al., 1981, Shaunessey et al., 1979; 1981,; Simpson et al., 1981). We

**Fig. 3.** View of Console 1 from the opposite side as that in Fig. 2. In the foreground are 4 pipette tips which will dispense cells in suspension on the microscope slides. To the left is the nebulizer which exudes mild steam on the slides after the cell suspension is dropped on them. The slides pass over a warming plate.

redesigned and rebuilt Console 2, installed a new barcode system, and wrote entirely new software. The system then operated routinely as part of our clinical service from 1981–84 before it was dismantled in 1985 due to increasing maintenance needs generated by aging components.

We describe the structure and operation of CASPS, our methods for evaluating the quality of slides produced by the system, and our procedure for trouble-shooting in order to improve the system at Northwestern. These methods should be of value in evaluating and developing other systems. We also present criteria and suggestions for the construction of new systems. As Priest (1977) said regarding variables affecting cell culture, "It becomes obvious that many procedural modifications are allowable; the problem is always to decide which ones are important."

## 2. The specimen preparation system

### 2.1 Description

The specimen preparation system consisted of 2 consoles under control of a PDP 11/34 minicomputer, a cathode ray tube and a printer. One limitation was that the consoles took up a lot of space (Fig. 1).

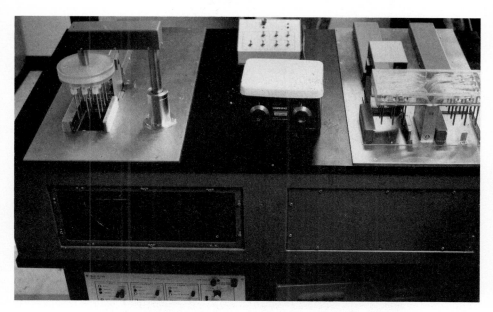

**Fig. 4.** Aerial view of Console 2. To the left is the aspirator (12 metal tubes, enough for 1 tray). In the center is the mixer. Trays were placed on the mixer to activate the magnetic stirrers in each culture well. Mixing also takes place on the fluid dispenser table. To the right are 2 sets of fluid dispensers, each with 12 metal tubes. Containers for fluids were placed inside the Console to the right.

Console 1 was used for culture initiation. Samples of blood were transported from 3ml vacutainers to trays of 12 culture wells (Fig. 2) containing nutrient media supplemented with fetal calf serum, antibiotics, and hepes buffer. This Console was also used at the end of the entire process to "drop" cells in suspension onto microscope slides (Fig. 3). The major components of Console 1 were a) three motorized tables for positioning samples and trays (Fig. 2 and 3); b) a rotating carousel which picked up and dispensed blood samples or cells in suspension (Fig. 2 and 3); plastic pipette tips which were automatically removed after a single use; c) a 4 channel slide conveyor with a nebulizer and slide warmer (Fig. 3); d) a bar code reader (bar codes with patient identification numbers were on vacutainers, culture trays and slides). Console I was enclosed in a plastic cover to minimize risk of contamination (Fig. 1). An ultraviolet light was put on overnight. At Console 2, supernatants were aspirated and solutions (hypotonic, fixative) were added (Fig.4). Microscope slides were stained by a Gam Rad automated slide stainer (Fig. 5).

After transport from the City of Hope, the system was reassembled at Northwestern. Initial runs indicated that slide quality was too poor for clinical analysis (few, clumped metaphases, intact nuclear membranes). There were also mechanical problems inhibiting routine operation. The manual prepara-

**Fig. 5.** The Gam Rad Stainer. 5 plastic dishes are in a circle to contain fluids for staining e.g. trypsin, distilled water for washes, Giemsa. In the foreground is the arm holding the trays with slides were prepared on Console 1 and placed on the Gam Rad in batches of 48. Control dials are in the foreground. Once set, the arm automatically moved the slides to each plastic dish and dipped them in for whatever times were set by the operator.

tion needed to prime the pumps on Console 2 daily was tedious and time consuming. The bar code identification error rate (20%) was unacceptably high. Furthermore, the software was neither "user friendly" nor easy to change, being written in 5 separate programs in assembler language.

## 2.2. Operation of the system

Blood specimens were obtained either directly in 3ml heparinized vacutainers, each of which had a bar code label on the side and a silver dot on the bottom, or were transferred into unheparinized vacutainers from a heparinized syringe. In the original City of Hope system a light pen was used to check patient identification by manually scanning a bar code on each vacutainer and placing each vacutainer in a metal matrix which recorded vacutainer entry by reflection off the silver dot. To save time this entire process was omitted in the Northwestern software, where double entry of the bar code number in the keyboard was substituted. Vacutainer numbers were added to the patient data file for accurate patient identification. This saved time by omitting placing silver dots on each vacutainer, and using the light pen. For many of our matched experiments (Section 2.3), blood was drawn into a syringe, dispensed into the vacutainers, and the remainder in the syringe processed manually. Intermittent inverting of the syringe was performed to avoid bias in allocating cells to the machine or manual methods. Blood was dispensed as simultaneously as feasible.

Forty eight vacutainers were placed in a rack which was rotated on a mechanical device to mix the blood. The rubber vacutainer tops were removed manually under a laminar flow hood, and a large piece of parafilm was placed on top to avoid contamination in transit to Console I. The rack was then placed on station one (Fig. 2). Media was dispensed manually to the 48 culture wells (12 wells/tray). A bar code was on the side of each tray. 48 disposable plastic pipette tips were placed on Station 4. The 4 trays were placed in a metal magazine and placed on Station 2 of Console 1. The culture intitation program was started. The rotating arm (carousel) on Console 1 then picked up pipette tips at Station 4 (4 at a time), picked up blood from four vacutainers at Station 1, and dispensed blood into 4 culture wells (from one vacutainer to one well) (Fig. 6). A mechanical arm moved each tray out under computer control. The carousel than dropped off the original 4 pipette tips and picked up a new set. This process continued until all 48 vacutainers were sampled. Provision to resample from the same vacutainer into a second or third set of 4 culture trays was incorporated into the new software. This procedure was used when large numbers of metaphases were required for radiation dosimetry or fragile X studies. In practice, 3 cultures were easily set up from each 3ml vacutainer.

Culture wells (4 trays) were then put in plastic transfer cases and incubated in routine fashion at 37°C. The plastic case was required to prevent contamination when carrying wells from the system to the incubator, and to protect the uncovered wells in the open system (5% $CO_2$). Ph monitoring indicated the gas entered the plastic cases adequately. After incubation, culture wells were centrifuged. A special centrifuge had the capability of spinning two entire magazines each holding 4 trays of 12 wells each. Two magazines were then placed on Console 2. An aspirator with 12 metal tubes removed supernatant from each tray. The aspirator head was sonically cleaned in between each tray aspiration. The magazines were then moved manually to the other side of Console 2. The hypotonic solution was prepared in a large container (enough for 48 wells) and placed inside Console 2. At some time before the next step, stir bars were placed in each culture well. Hypotonic solution was then added dropwise (Fig. 4).

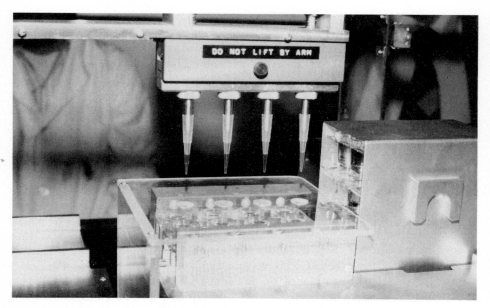

**Fig. 6.** Pipette tips poised above 4 culture wells ready to dispense blood.

The magazines were then returned to the incubator for a certain time. A device in Console 2 caused the stir bars to rotate after each hypotonic and fixative addition to the wells. Aspiration was repeated after each centrifugation. Fixative was dispensed to one tray (12 wells) at a time. (Fig. 4). After redevelopment of Console 2 (Section 2.4), two large separate wells contained the fix components – methanol and acetic acid. These components were automatically mixed in proportions specified by the operator through the computer program. The wells were temperature controlled. In our protocol the hypotonic solution was warmed; the fixative chilled. Fixation, storage, and subsequent centrifugation were repeated as many times as desired.

The final step occurred on Console 1. Cells in suspension were aspirated from the culture wells by the disposable pipette tips and dropped on slides with bar codes. The leading edges of the slides were beveled to avoid slide jams. The slides were moved in parallel 4 at a time, under the tips. Bar code I.D.'s were stored in patient records. Slides were then passed over a warming tray, exposed to a mild steam from a nebulizer, stacked in four magazines and stained in the Gam Rad. Two slides, each with a bar code, were prepared from each culture well.

## 2.3. Determination of slide quality: automated vs. manual

Many parameters are involved in the preparation of microscope slides for cytogenetic analysis from human lymphocyte cultures. Variations exists in each step in the production of slides: (1) specimen collection (collection vessel, anticoagulant, storage), (2) culture initiation (sample volume, medium composition, environment), (3) incubation (culture volume and vessel, pH, duration, temperature), (4) harvest (centrifugation speed, agitation, duration of hypotonic, fixation speed, centrifuge speed, fixative components, temperature and freshness of fixative, and number of repetitions), (5) slide dropping (method of slide cleaning, wetness, dropping height), and, ultimately, staining. Humidity of the environment in which slides are made is also a factor. Individual cytogeneticists tend to vary several factors simutaneously to obtain and maintain good slide quality. The resultant variety of techniques, differing from one another in sometimes small, and sometimes large details, is truly vast and oftentimes poorly documented, with no indication of which of several possible parameters is producing the desired effect. Thus the question arises concerning the true importance to slide quality of each of the details of these protocols (Priest, 1977).

We believed that evaluation of an automated system required systematic, blind, objective procedures to differentiate inherent biological and laboratory variation from effects of automation to identify the effects of single variables as well as interactions, and to minimize personal preference and whims of cytogeneticists in determining slide quality. The series of blind, controlled trials we conducted were similar in that the results of all experiments included a blind comparison of the quality of the output (microscope slides) obtainable by routine manual methods with that from matched blood specimens processed by CASPS. In our preliminary trials, for slide quality, the number of cells, number of metaphases, and metaphase quality on a 1–4 scale were noted. In later trials to "fine tune" the system, we not only counted the number of metaphases on the slide, and determined the mitotic index, but also recorded the percent of those the technologist considered analyzable based on an objective plus subjective rating (Appendix A). Comparisons for slide quality were most reproducible the more objective the scoring methods. When cytogeneticists used a numerical subjective evaluation of slide quality, the overall agreement among raters (cytogenticists) was not good, as measured by a weighted Kappa statistic (Aivano et.al, 1976; Cichetti, 1976; Cichetti et al., 1977). For perfect agreement, K=1; our results for 4 cytogeneticists using subjective slide quality evaluation were K=0.46. Raters agreed best on "fair". They also tended to reevaluate the same slide as "fair" when presented to them again without their knowledge, regardless of how they scored the slides originally. That is, raters tended to give different scores to the same slides scanned at different times. Finally, we determined that for *metaphase* quality part 1 of Appendix A (Distribution of Chromosomes) together with determining how many bands were visible in chromosome No. 18, was a reliable metaphase rating method. For our final clinical trials after redesigning CASPS and deciding on a final protocol, a "diagnostic unit" was considered to be chromosomes counted in 5 metaphases, and one karyotype prepared; or 20 metaphases and two karyotypes.

The statistical methods used which were developed by one of us (AD) for this project are shown in Appendix B. Slides were analyzed independently by several cytogeneticists.

There were 3 sets of trials:

1. initial trials to plan hardware redesign and protocol changes to improve the number and quality of metaphases;
2. trials aimed at fine tuning of the system to improve banding quality after rebuilding Console 2;
3. clinical trials to determine the adequacy of slides produced by automation for diagnosis.

   (For reasons of space, only part of the data from these trials is presented in this paper. More information is available upon request from the authors.)

Finally, we assessed other factors related to the overall acceptance of the system as a clinical and research tool e.g. cost-benefit (Cimino et al., 1982).

**2.3.1. Initial trials (tables 1a and b).** In order to determine the step(s) responsible for the poor slide quality observed initially, factors involved in the culture initiation and harvesting process were listed (Table 1A and B). Some of these variables were expected *a priori* to affect mitotic index and metaphase quality. Each variable was tested separately. Items with an asterisk in Tables 1A and B were found to affect slide quality. The greatest improvement was effected by increasing the amount of blood dispensed per culture well (Trial 23679, 0.2–0.25ml blood/3ml well was optimum).

As a result of these initial trials, a protocol was developed which provided enough metaphases on each slide to proceed with the next set of trials.

**Table 1.** Individual steps in specimen preparation to be evaluated

### A: Culture initiation variables

1. Relative amount of blood and media*
2. Type of heparin
3. Amount of phytohemoglutinin*
4. Exogenous $CO_2$
5. Sterile techniques
6. Incubation
   A. Time*
   B. Temperature
7. Ph*
8. % Fetal calf
9. Type of media*
10. Interval between sampling and culture initiation

*Indicates those with demonstrable effects on slide quality based on comparisons of matched samples processed by CASPS versus routine manual methods.

## B: Culture harvesting variables

1. Colchicine
   A. Type (colchicine vs. colcemid)
   B. Amount (concentration)
   C. Length of exposure
2. Hypotonic
   A. Type
   B. Concentration
   C. Time
   D. Temperature
   E. Mechanical agitation
   F. Relative volume
   G. Number of changes
3. Centrifugation speed and time* (used after all harvesting steps)
4. Fixation
   A. Time
   B. Type (composition, quality, "freshness")
   C. Mechanical agitation
   D. Temperature
   E. Number of changes
   F. Relative volume
   G. Final cell concentration
5. Preparation of slides
   A. Cleanliness of glass
   B. Temperature of glass and fix
   C. Height from which cell suspension dropped on slide
   D. Number of drops
   E. Concentration of cells
   F. Time of heating slide
   G. Temperature of heating surface
   H. Temperature and amount of steam
   I. Humidity*
6. Staining
   A. Type of staining concentration
   B. Amount and time
   C. Pretreatment e.g. trypsin
   D. Ph
   E. Age of slide
   F. Type of storage prior to staining

**2.3.2. Determining the part of the automated process causing slides to be of lesser quality than those prepared manually.** We wished to identify which part of the automated process was lowering slide quality. Effects of 6 key steps for slide production were examined separately (Appendix C). Each step was tested independently holding the other steps constant. Each blood specimen was split. On one half of the specimen, all steps were performed manually except for *one* step performed on the machine: conversely all steps were performed on the machine except *one* step performed manually. An overall comparison of the culture volumes used routinely (3ml CASPS, 6ml manual) were trials 1 and 2. Other matched trials were run using 3ml when all but one step was on CASPS, 6ml when all but one step was performed manually. Slides were then blindly evaluated by use of a standard coding sheet. Because the bar codes would identify the automated slides, bar codes were also placed on slides prepared by manual methods.

We found that the poorest slides among those where all steps were performed manually except one, resulted from the experiments where fixation or the slide dropping steps were on CASPS.

Conversely, the best slides when all but one step was done on CASPS, were those with manual fixation or slide dropping. This identified the problem in CASPS as fixation and slide dropping. The Console 1 slide dropping problem was not correctable due to inherent limitations of the machine – slide "dropping" height was too low, the amount of cells in suspension dropped on each slide was too small (0.25ml), and the warming and nebulizing occurred after cells were dropped on the slide. These problems could be corrected in a second generation machine.

However, we were able to rebuild Console 2 to solve the fixation problem (Section 2.4). In a series of 50 trials (a sample of which are shown in Table 2) using matched blood specimens, we then "fine tuned" the system by varying single parameters to see if we could further improve slide quality by altering components of processing, rather than the system itself. We incorporated the best variables into our final protocol. The most significant variables were blood source, and 96 hr. incubation.

**Table 2.** Summary of results: trials on effects of parameters in specimen preparation on slide quality. Unless noted, statistical methods are described in Appendix B.

| Trial Date | Null Hypothesis | Endpoint | Results |
|---|---|---|---|
| **FIXATIVE CONCENTRATION** | | | |
| 09881 | No difference in 5:2 vs. 1:1 fixative (methonol: glacial acetic acid) | No. of metaphases | 5:2 better than 1:1 |
| 11781 | No difference in 5:2 vs. 1:1 fixative (methanol : glacial acetic acid) | No. of metaphases<br><br>% Analyzable metaphases | No difference<br><br>1:1 better than 5:2 |
| 13181 | No effect of Gibco vs. Difco PHA and 5:2 vs. 1:1 fix (interaction) | No. of metaphases<br><br><br>% Analyzable | Difco better than Gibco for number of metaphases;<br>no difference % analyzable (slides prepared in the afternoon are better than those prepared in the morning) |
| 12681 | No difference in 6:1 vs. 2:1 fixative | No. of metaphases<br><br>% Analyzable | 6:1 better than 2:1<br><br>6:1 better than 2:1 |
| 16181 | No difference in 6:1 vs. 1:1 fixative | No. of metaphases<br><br>% Analyzable | 6:1 better than 1:1<br><br>No difference |
| 18281 | No difference in either hypotonic time (10 vs. 30 min) or fix concentration (5:2 vs. 6:1) | No. of metaphases<br><br>% Analyzable | 6:1 fix best<br><br>Not analyzable<br><br>Interactions 5:2 and 30 min; 6:1 and 10 min best |
| **HEPES BUFFER** | | | |
| 25981 | No difference when cultures are grown with vs. without hepes buffer | No. of metaphases<br><br>% Analyzable | No difference<br><br>Better with hepes |

| Trial Date | Null Hypothesis | Endpoint | Results |
|---|---|---|---|
| | **PHYTOHEMAGGLUTININ TYPE** | | |
| 13181 | — see above — | — | — |
| 14081 | No difference in Gibco vs. Difco PHA and/or time in hypotonic (25 vs. 10 mins) | No. of metaphases | Difco better than Gibco<br><br>No difference 25 vs. 10 min. |
| | **HYPOTONIC TIMES** | | |
| 14081 | — see above — | — | — |
| 16881 | No difference in hypotonic time (25 vs. 10 min) | No. of metaphases<br><br>% Analyzable | No difference<br><br>No difference |
| 17581 | No difference in hypotonic time (15 vs. 30 min.) | No. of metaphases<br><br>% Analyzable | No difference (AM better than PM)<br><br>No difference |
| 18281 | — see above — | — | — |
| | **INCUBATION TIMES** | | |
| 15581 | No difference in 72 vs. 96 hr. incubation time | No. of metaphases | 96 hrs. better |
| 19681 | No difference in 72 vs. 96 hr. incubation time | No. of metaphases<br><br>% Analyzable | 96 hrs better<br><br>No difference |
| | **COLCHICINE TIME** | | |
| 18981 | No difference in either colchicine time (30 vs. 60 min) or additional stirring after aspiration | No. of metaphases metaphases<br><br>% Analyzable | No difference for colchcine time; no for stirring<br><br>No difference |
| | **CENTRIFUGE TIME AND SPEED** | | |
| 20181 | No difference in either centrifuge time (5 vs. 10 min) or speed (1,000rpm, 15,000rpm, 2,100rpm) | No. of metaphases<br><br>% Analyzable | No difference, although 2000 rpm at 10 min. best |
| | **ADDED MECHANICAL AGITATION** | | |
| 18981 | — see above — | — | — |

| Trial Date | Null Hypothesis | Endpoint | Results |
|---|---|---|---|
| **SEQUENTIAL SAMPLING OF CULTURE WELLS** | | | |
| 27381 | No difference in slides made from same culture wells | No. of metaphases | No difference |
| | at sequential sampling intervals | % Analyzable | No difference (Freedman's Rank Test) |
| **PROCESSING IN MORNING VS. AFTERNOON** | | | |
| 14781 | No difference between bloods processed in morning (A.M.) | No. of metaphases | A.M. better than P.M. |
| | vs. afternoon (P.M.) | % Analyzable | No difference |
| **SLIDE DROPPING, PLASTIC VS. GLASS PIPETTES** | | | |
| 26681 | No difference when cells in suspension are dropped onto | No. of metaphases | No difference |
| | slides by automated vs. manual methods | % Analyzable | Manual better than automated |
| 30781 | No difference using plastic vs. glass pipettes to drop slides; no difference manual vs. automated | No. of metaphases | Manual dropping with plastic or glass pipettes better than automated dropping with plastic; |
| | slide dropping | % analyzable | No difference glass vs. plastic |
| **OVERALL – AUTOMATED VS. MANUAL** | | | |
| 24381 | No difference when automated vs. | No. of metaphases | Less with automated |
| | manual methods | % Analyzable | Less with automated |
| | used to prepare slides | Mitotic index | No difference |
| 21181 | No difference in karyotypes prepared from slides made by automated system compared to those prepared in routine fashion | Ranks of 2 cytogenticists on 54 pairs (108 karyotypes) | No difference |

**2.3.3. Clinical trials.** The final set of clinical trials using the optimum protocols indicated that slide quality was comparable to that prepared manually as judged by obtaining accurate diagnosis (chromosomes in 5 metaphases counted and one karyotype prepared per slide, or 20 metaphases and 2 karyotypes). No contamination was ever observed. The *number* of metaphases per slide was always less when produced by the automated system, but there were still sufficient metaphases to achieve the number required for diagnosis. Four slides from the automated system were required on the average (this is equivalent to 2 slides from routine processing due to half the automated slide being covered with a bar code label). Interestingly, backgrounds appeared cleaner on automated slides, an original goal of Castleman and Melnyk.

A service grant from the March of Dimes allowed us to offer cytogenetic analyses free of charge to clinic patients who had experienced more than one spontaneous abortion, and to survey institutions for the retarded. When we tested the system on blood from known fragile X males, 4 wells were required per patient (8 slides) to reliably obtain 100 analyzable cells. If large numbers of metaphases were required per patient e.g. for radiation dosimetry, more culture wells per patient might have had to be initiated. We programmed the machine to resample from the 3ml blood vacutainer to prepare several culture wells rather than the one to one in the original system. This permitted many slides to be made from each blood specimen. More than one 3ml vacutainer per patient could also be initiated, but we did not find that to be necessary to obtain 20–100 metaphases. This might be necessary for the 48 hr. incubation periods necessary for radiation dosimetry studies.

Karyotypes were prepared from slides prepared from matched samples on CASPS and by manual methods from all these clinical samples. The forms were coded, randomized in pairs, and presented to cytogeneticists to select one of each pair for diagnosis. The automated karyotype was selected as often as the manual one (Sign Test $P < .05$).

## 2.4. Increased blood volume dispenser; fluid dispenser redesign; new bar code system

Our conclusions from results of the initial experiments were that the problems causing poor slide quality occurred in Console 2 during the fixation steps, and in Console 1 during slide "dropping." Console 1 was modified to dispense 0.17ml of blood rather than 0.08ml. We were unable to increase the amount to the 0.20–0.25ml optimum. We corrected the Console 2 problem by installing a new pump system and rebuilding that area. We were not able to correct the slide dropping problem due to mechanical limitations inherent in Console 1, i.e. the height from which cells in suspension were dropped could not be increased.

The original fluid dispenser installed in the CASPS system consisted of 24 peristaltic pump lines driven by four pump heads. Problems encountered with the original pump system fell into several categories; operational, functional, and maintenance. Numerous difficulties were encountered in the day to day operation of the fluid dispenser. The most notable was the excessive amount of time required to set-up and prime the system. Each channel had to be individually clamped prior to priming. This daily reclamping was required to preserve the already short life of the tubing used in the pumps. Once the tubing has been clamped, the pumps had to be primed with old solutions to ensure operation without wasting fresh chemicals. When proper operation was verified, fresh fluids were then introduced into the system. The entire pump preparation took approximately 15–25 minutes, 20% of the time required for an entire automated harvest! The tubing for fixative had to be changed every 10 harvests and the hypotonic tubing changed every 20. Approximately 45 minutes is required to change one set of tubes.

The capability of changing flow rates at the already low volume, and the corrosive nature of the fluids used, called for a novel arrangement. The new pumps we chose (IVEK Corporation) were a piston type operated by stepping motors to achieve a high degree of control over the flow rate. Flow rate with such pumps can be set and changed within a cycle, that is, they have the capability of changing almost instantaneously. The pistons and cylinders of these pumps are manufactured of a ceramic material which is impervious to the chemicals employed.

The system was capable of receiving 5 volt signals to start operation, and of sending cycle complete signals. This feature made interfacing this system with the existing hardware quite easy. It was capable of mixing the components of fixative just prior to their use. This eliminated a significant variable in the harvest procedure (fixative freshness). Fluids were distributed to the individual wells via a manifold with valves to make up for any variations that might be encountered in the flow rate.

The modified fluid dispenser required less than 2 minutes to prime. (compared to 25 minutes in the previous system). There were no operator actions required once the system has been primed (as opposed to the unclamping in the peristaltic pumps of the old system) and there were no periodic maintenance requirements outside of replacement of failed parts (pump heads should be replaced every 2000 hrs.).

The original bar code reader had too many errors for clinical diagnostic use. An Intermec Bar Code system replaced the old bar code system. This change reduced errors to low, clinically acceptable levels.

## 2.5. Software

A new software package for operation of the Automated Specimen Preparation system was written. This was a "user friendly" system to operate the console and to store patient information.

The Northwestern software package (CASPS) consists of a main Fortran IV program and several subroutines in Macro 11 assembler language which drove the individual hardware modules. The package ran under the RT-11 Version 3 Operating System and, due to its reliance on subroutines, provided a more structured approach and was more easily understood and updated than the software previously in use at City of Hope.

In addition to running and controlling the system hardware, the program provided the means for creating and editing data files in which patient and experimental data were stored. An additional program, SUMM, was written by HS. This program reduced the information contained in the master data file created by CASPS and placed it on a summary sheet for the laboratory notebook. Documentation of this software is available from the senior author upon request.

## 2.6. Cost-benefit analysis

The automated system achieved significant laboratory efficiency in two phases, harvesting specimens and staining slides (Cimino et al. 1982). 220 cultures/day could be processed with CASPS, compared to 30–54 cultures/day manually harvested if one techician worked as fast as possible. This was a favorable 4:1 ratio.

The Gam Rad stained slides (trypsin-Giemsa) in batches of 24. This machine was mechanical, not computerized. An arm dipped trays holding 24 slides into a sequence of plastic dishes in which solutions were placed. The times/station was set by the operator. The Gam Rad saved large amounts of time; 4 times as many slides could be processed per day (960 vs. 240 manual, a 4:1 ratio).

Based on a trial consisting of 21 batches of 48 cultures each, the average time to initiate 48 blood specimen culture was 10 min after the vacutainers were placed on Station 1. Including computerized data entry total time was 40 minutes (less than 1 min./culture). A technician could manually dispense bloods in culture vessels in that time, so automation provided little advantage. The number of cultures which could be initiated by one technician doing nothing else in an 8 hour day is 535 with CASPS compared to 288 manually (Cimino et al., 1982). This is not a large difference (2:1) because both methods require making up media, dispensing it into tubes or wells and preparing and labelling tubes. A separate media dispenser had been used at the City of Hope, but because it required large amounts of media and did not save time, it was not used at Northwestern. Manual methods for patients' data entry were not comparable to that for the computer because the equivalent manual methods included both secretarial and technician recording, whereas using CASPS one operator did everything. However, subjectively, an advantage of the automated system was noted in that both were combined and shortened.

Comparison of labor and supply costs using our final protocol (Cimino et al., 1982) indicated that overall labor costs were dramatically reduced by 64% compared to manual processing. Labor costs for 48 cultures totaled $101.30

using manual methods, $28.70 using CASPS assuming wages and overhead of $10/hr. One person – *not* necessarily a cytogeneticist – could process more than 12,500 cultures a year (260 working days) running *only* 1 batch a day. Two batches could be run increasing output to 25,000 cultures per year. Few systems of this type would be required in the entire world, unless large scale cytogenetic analysis including perhaps environmental monitoring were desired. The analysis phase would require another 50% effort. Automated metaphase locator and karyotype devices would be required to keep up with the rapid slide production. If it becomes possible to detect and analyze fetal cells in the maternal circulation, some automation may be needed to process large numbers of specimens.

## 3. Second generation systems

For a second generation system we propose a single unit, desk top version under microprocessor control, performing both "harvesting" and slide dropping. We determined that 6 cultures per batch was the minimum number at which automation showed advantages in time over manual processing. Generic rather than special purpose components, such as the wells on CASPS, are preferred for ease of supply availability. Culture initiation of blood specimens did not show enough time savings to warrant automation, however, automated culture initiation for other specimens e.g. CVS direct preparations, is worth exploring at the present time. A media dispenser would save time. We anticipate that a laminar flow environment would be necessary to preserve sterility, and a different type of culture well (larger and with a flat bottom) would be required for prenatal specimens. Petri dishes holding 3ml and cover slips are popular ways to grow amniotic fluids and chorionic villus specimens, so this type of "well" might be considered for automation. In the TECAN (Spureck et al, 1988) blood cultures are harvested in chamber slides.

The proposed device should consist of a single, reasonable sized console which will have the following capabilities:

1) aspirate supernatants
2) add fluids mixed at various temperatures and concentrations
3) centrifuge
4) maintain a sterile environment for media changing
5) have microprocessor control
6) "drop" slides

Cost of the systems should be kept moderate, in the $20,000–40,000 range.

## 4. Conclusion

In this paper we have described our experience with an automated specimen preparation system in a clinical setting. Initial evaluation indicated the need for further development. After redesign of the pump system of CASPS at Northwestern, implementation of "user friendly" software, and a new bar code reader, our results show that an automated specimen preparation can operate successfully in a clinical environment. Extremely large numbers of specimens could be processed with one, cytogenetically *inexperienced* operator, saving trained personnel for analysis. Increased speed and batch processing yielded favorable cost-benefit ratios. Having demonstrated the value and feasibility of automated specimen preparation, at least for bloods, we recommend pursuit of second generation machines which can take advantage of current technology and be used for all types of specimens.

Based on results of our experiments, we have identified the specific major factors affecting quality of blood specimen preparation to be 1) blood source, different persons showing large variation in mitotic index; 2) 96 hr incubation, which yielded the best mitotic index; 3) brand of phytohemagluttinen (PHA) (Difco was the best); 4) blood volume (0.17ml blood/2ml media); 5) slow addition of cold, fresh fixative; 6) dropping cells from a 10 inch height onto wet slides.

Although laboratory cytogeneticists will not be surprised at these results, our studies indicate that objective determinations of these factors should be the prime foci in developing automated systems.

*Acknowledgements.* Supported in Part by NIH N01-HD-8-2855, and a March of Dimes Service Grant. Assistance of Ms. Patricia Lathion in preparation of this manuscript is greatly appreciated.

# Appendix A - Scoring sheet for evaluation of slide quality

A. Distribution of chromosomes
  1. No overlapping or touching chromosomes and well spread i.e. neither too scattered nor too close
  2. No overlapping or touching chromosomes but scattered over too many fields for ease of counting.
  3. Some overlaps and /or touching present but karyotypable (relative to contraction state, etc.
  4. Too many overlaps and/touching chromosomes to be good for karyotype but still countable.
  5. Too many overlaps and/or touching chromosomes to permit accurate counting i.e. not analyzable (also including "too clumped")
B. Contraction state of chromosomes:
  1. Interphase
  2. Prophase
  3. Prometaphase
  4. Metaphase - elongated
  5. Metaphase - average length
  6. Metaphase - contracted
  7. Other - e.g. not distinguishable
C. Banding pattern (realtive to the contraction state, i.e. relative to the best that can be expected at that stage):
  1. Needs no (or very little)improvement
  2. Needs improvement
  If you checked C1, skip to section G; otherwise, continue
D. Stain quality
  1. Staining too dark i.e. obscures bands
  2. Staining too light i.e. not good for photo
  3. Staining correct intensity
E. Trypsinization:
  1. Overtrypsinized
  2. Undertrysinized
  3. Trypsinization correct
F. Chromosome morphology:
  1. Poor — uncertain cause
  2. Too bent and /or twisted for band pattern analysis
  3. Adequate
G. Nuclear membrane:
  1. Intact
  2. Not discernible
  3. Diffuse
H. Poor background or major artifacts, e.g. chromosomes obscured by interphase nucleus (ignore minor variations which don't interfere with the analysis)
  1. Present
  2. Absent
I. Other problems
  1. Absent
  2. Present (specify)
J. Subjective rating:
  1. Excellent
  2. Very good
  3. Fair
  4. Poor

# Appendix B - Analysis of experimental results

**A.** The analysis of parameters used in blood specimen preparation has been done using a variation of the two-sample rank test (non-parametric) derived by AD. Details of the analysis for one type of trial are presented.

**Example:** For each blood on each day there are two slides from the morning run (AM) and two slides from the afternoon run (PM), or two slides each for treatment A and treatment B. Rank the outcome variable corresponding to this experiment for the four slides from 1 to 4, i.e., the slide with the largest value receives a 4, the next largest 3, and so on. Sum the ranks corresponding to treatment A and treatment B, and take the difference, i.e.,

$$R_{A1} + R_{A2} - R_{B1} - R_{B2}$$

This difference ranges from +4 to -4, and has an expected value of 0, under the null hypothesis of no treatment difference.

Define $D_{i1}, i = 1, ..., n$, to be the difference in the sum of the ranks for the $i$th blood on day one.

Then under H$_0$,

$$E(D_{i1}) = 0, Var(D_{i1}) = 20/3$$

To compare the treatments for six bloods on one day, sum the $D_{i1}$ for the bloods. The test statistic is then

$$T = \frac{\sum_{i=1}^{6} D_{i1}}{\sqrt{\sum_{i=1}^{6} Var(D_{i1})}}$$
$$= \frac{\sum_{i=1}^{6} D_{i1}}{\sqrt{6 \cdot 20/3}}$$
$$= \frac{\sum_{i=1}^{6} D_{i1}}{\sqrt{40}}$$

Compare this to a normal distribution. This sum can only take on values in multiples of 2.

| $\left\vert \sum_{i=1}^{6} D_{i1} \right\vert$ | P |
|---|---|
| 12 | .0574 |
| 14 | .0272 |
| 16 | .0114 |
| 18 | .0044 |
| 20 | .0016 |
| 22 | <.0010 |

**B.** This test can also be done by summing only the ranks for treatment A, i.e. $R_{A1} + R_{A2}$.

Let $S_{i1} = R_{A1} + R_{A2}$. To compare all 12 bloods on both days, sum all 12 D's or all 12 S's.

$$T_2 = \frac{\sum_{i=1}^{6} D_{i1} + \sum_{i=1}^{6} D_{i2}}{\sqrt{12 \cdot 20/3}}$$

$$= \frac{\sum_{i=1}^{6} D_{i1} + \sum_{i=1}^{6} D_{i2}}{\sqrt{80}}$$

The significance of the numerator of T2 is:

| Numerator | P |
|-----------|---------|
| 16 | .0754 |
| 18 | .0444 |
| 20 | .0251 |
| 22 | .0139 |
| 24 | .0074 |
| 26 | .0036 |
| 28 | .0017 |
| 30 | <.0010 |

One can also define a test statistic by adding $\sum_{i=1}^{6} S_{i1} + \sum_{i=1}^{6} S_{i2}$, i.e., the sum of all ranks corresponding to treatment A. Under $H_0$,

$$E(S_{i1}) = 5$$
$$\text{Var}(S_{i1}) = 5/3$$

To compare treatment A and B on six bloods,

$$T_1 = \frac{\sum_{i=1}^{6} S_{i1} - 30}{\sqrt{\sum_{i=1}^{6} Var(S_{i1})}}$$

$$= \frac{\sum_{i=1}^{6} S_{i1} - 30}{\sqrt{10}}$$

| $\sum_{i=1}^{6} S_{i1}$ | P |
|-----------|---------|
| 36 | .0574 |
| 37 | .0272 |
| 38 | .0114 |
| 39 | .0440 |
| 40 | .0016 |
| 41 | <.0010 |

**C.** For some experiments, we defined a sign test, i.e., if on a given day, both reviewers had larger values for treatment A than treatment B a + was assigned, while if one is larger but the other smaller, that is a tie, and if both are smaller, a –.

The hypothesis of no difference, says plusses and minuses are equally distributed. This is not as powerful as the rank test, but it is easier to do.

## Appendix C - Trial 35279

Single component subtrials to test the effects of individual steps of the automated specimen preparation system upon slide quality

| Design | | | Steps | | | |
|--------|--------|--------|--------|--------|--------|--------|
| Subtrial number | Culture initiation | Culture volume (ml.) | incubation | hypotonic | fixative | Slide dropping |
| 1 | A | 6 | M | | | |
| 2 | A | 3 | M | | | |
| 3 | M | 6 | A | M | | |
| 4 | M | 6 | M | A | M | |
| 5 | M | 6 | M | | A | M |
| 6 | M | 6 | M | | | A |
| 7 | M | 6 | A | | | |
| 8 | M | 3 | A | | | |
| 9 | A | 3 | M | A | | |
| 10 | A | 3 | A | M | A | |
| 11 | A | 3 | A | | M | A |
| 12 | A | 3 | A | | | M |

Code:   A  = automated procedure
        M  = manual procedure

## References

1    Aivano SL, Cicchetti DV, Levine J. Selecting the most reliable from a set of judges. Proc. Amer. Stat. Ass., Social Stat. Sect. pp.145-149 (1976)
2    Castleman KR, Melnyk JH. Final Report. NIH N01-HD-5-2843 (1975)
3    Cichetti DV. Assessing inter-rater reliability for rating scales. Resolving some basic issues. Brit. J. Psychat 129:452-456 (1976)
4    Cicchetti DV. Aivano SL, Vitale J. Computer programs for assessing rater agreement and rater bias for qualification data. Educ. and Psychol. Measurement 37:195-201 (1977)
5    Cimino MC, Martin AO, Maremont SM, Shaunnessey MD, Simpson JL. Automated cytogenetic processing for human population monitoring. In: Bora DC (ed) Chemical Mutagenesis, Human Population, Monitoring, and Genetic Risk Assessment (Elsevier Northholland Biomedical Press, Amsterdam, 1982) pp.169-173.
6    Lundsteen C, Martin, AO. On the Selection of Automated System.... Amer. J. Med. Genet. in press (1989)
7    Martin AO, Unger, N. Automated cytogenetics, Comtemporary Ob-Gyn Technology (1985)
8    Martin AO, Shaunnessey M, Sabrin H, Maremont M, McKinney RD, Cohen MM, Rissman A, Kowal D, Cimino M. Automation of cytogenetic preparation and staining techniques; Report of European Work Group on Automated Chromosome Analysis, Berlin, (1986)

9      Martin AO. Final Report. NIH NO1-HD-5-2855 (1981)

10     Martin AO. My life with two automated systems. Preliminary report on work group meeting an Automated Chromosome Analysis, Leiden, Netherlands, Oct. 17-19, pp 52-74 (1985)

11     Melnyk JH, Persinger G, Mount B, Castleman KR. A semi-automated specimen preparation system for cytogenetics. J. Reprod, Med. 17:59-67 (1970)

12     Priest JT. Medical Cytogenetics and Cell Culture. (Febrigee, Philadelphia, 1977) p.142.

13     Sabrin HW, Martin AO, Shaunnessey MS. Current status of metaphase locating devices: Preliminary clinical evaluation of one system. In Proceedings of Computer Applications in Medical Care, Conference Nov. 1-4, , Washington, D.C., IEEE Computer Society Press Order, No. 377, pp.1100-1111 (1981)

14     Shaunessey M, Martin AO, Rzeszotarski MS, Thomas CW, Isenstein B. Status of a computer-aided system for cytogenetic specimen preparation and analysis. Proceedings of the Third International IEEE COMPSAC, pp. 279-281, (1979)

15     Shaunnessey M, Martin AO, Sabrin HN, Cimino MC, Rissman A. A new era for cytogenetic laboratories; automated specimen preparation. Proceedings of Computer Application in Medical Care, Conference Nov. 1-4 1981, Washington, D.C. IEEE Computer Seciety press Order No. 377, pp.538-541 (1981)

16     Simpson JL, Martin AO, Shaunnessey MD. Semi-automated chromosomal analysis: Current technological status and future prospects, Contemporary Ob-Gyn Technology pp.49-56 (1981)

17     Spureck J, Carlson R, Allen J, Dewald G. Culturing and robotic harvesting of bone marrow, lymph nodes, peripheral bloods, fibroblasts, solid tumors with in situ techniques. Cancer Genetics and Cytogenetics 32:59-66 (1988)

# Part V

# Automatic Chromosome Segmentation

# Decomposition of Overlapping Chromosomes

Liang Ji

## Summary

This paper is concerned with the general problem of finding plausible lines of separation between overlapping groups of dark objects. The particular problem domain considered is that of chromosome analysis, in which overlapping chromosomes cannot be separated by thresholding or by the methods for separation of touching chromosomes. We will show that the problem can be largely solved for the chromosome domain by a procedure based on analysis of concavities in relation to expected chromosome shape, and by use of the convex hull and the node of the skeleton of overlapping chromosomes. The individual chromosomes can be restored from the result of the analysis by using a linear least squares method. A 95% success rate was obtained for a set of 46 individual clusters of overlapping chromosomes.

## 1. Introduction

One of the major tasks in chromosome analysis automation is to obtain a good segmentation of the data. Although conventional thresholding can solve the major part of the segmentation problem, there are nevertheless still many exceptions. Currently most chromosome analysis systems rely on the operator to separate groups of touching and overlapping chromosomes which is both tedious and time consuming. Without a reliable method of automatic decomposition even a chromosome counter is very difficult to produce [19].

In recent years different methods for separation of touching clusters have appeared in the literature [3,15,22,23,11] but apart from the early papers [12,16], none of them have paid attention to the decomposition of overlapping clusters.

One reason may be that fewer overlapping clusters of chromosomes are found than touching ones. It is perhaps not an important problem unless a fully automatic chromosome analysis system is intended. Methods for splitting touching clusters cannot deal with overlaps. When two chromosomes are touching, there are cut points at which the two boundaries of the chromosomes intersect. In most of the decomposition methods the main task is to find the cut point pair [3,22,23]. Failures arise when the cut point pair, especially the second cut point, is chosen incorrectly [14]. In the case of a pair of overlapping

**Automation of Cytogenetics**   Editors: C. Lundsteen J. Piper
© Springer-Verlag Berlin Heidelberg New York 1989

chromosomes four cut points are involved, which must all be found correctly. In the next sections we will show why the conventional curvature analysis cannot meet the challenge of this task and introduce the method we have developed. Because only a few papers have appeared in the literature on this particular problem, we will describe in detail the consideration leading to the choice of our method. In the final two sections we give the result of testing and a short discussion. 46 overlapping groups of G-banded chromosomes were tested, and a 95% success rate was achieved.

## 2. Basic definitions and notation

The following notation or definitions are used throughout this paper.

*Object*: 8-connected subset of the discrete (x, y) plane, assumed to represent a chromosome or a pair of overlapping chromosomes, or occasionally a cluster of more than 2 overlapping chromosomes.

*Boundary* (of an object): set of points belonging to the object and 4-adjacent to points of the complement.

$p_i$ is the i-th vertex of the object's boundary, which is enumerated in anti-clockwise order.

*Cut point*: a point at which it is hypothesized that the boundaries of the overlapping objects intersect. A prospective split line starts and ends at cut points.

*Skeleton* (of an object): a "thinning" transformation of the object which is one pixel wide almost everywhere and preserves the connectedness and the shape of the original object. Many definitions of skeleton can be found in [18,21] and the reference papers. The detailed differences among the definitions are not important for our purpose.

*Node* (of a skeleton): a point in the skeleton which has at least three neighbours in the skeleton.

*End point* (of a skeleton): a point which has only one 8-connected neighbour.

*Branch* (of a skeleton): a sequence of 8-connected points which starts and ends at a node or end point but does not include any other nodes or end points.

We knew from the early work [16] that if four cut points are known the overlapped part of the boundaries can be reconstructed through these four cut points. If the cut points are known the rest of the decomposition procedure becomes relatively easy. So the key point in the decomposition of overlapping chromosomes is finding the cut points. In [16] the cut points are found by an operator and marked by a light pen, but how can they be found by machine? The following sections describe some methods which we have used. From their success and failure, the reasoning behind the choice of our present method will be easer to understand.

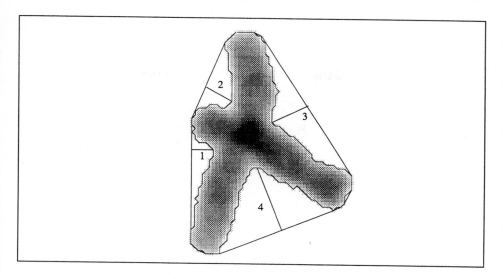

**Fig. 1.** Cut points found by deepest point method.

# 3. Procedures for finding cut points

## 3.1. Concavity analysis

This method can be outlined as follows.

   a) Construct a convex hull of the overlapping chromosomes.

   b) Find the four biggest connected regions between the boundary of the object and the chords of the convex hull. See fig 1, the regions labelled 1, 2, 3, 4.

   c) In each of the above regions search for a point on the boundary at which the distance between it and the corresponding chord is greatest. We term it the deepest point of the concavity. See the points in fig. 1 which are the intersection points of the boundary and the perpendicular line to the chords.

   d) These four points are the four cut points.

     This method worked well for a cross-like overlap (fig 1) but failed when one or both chromosomes are bent. The reason is that a cut point is not necessarily a "deepest" point in a concavity (see region 4 in fig 2); or more than one cut point may occur in the same region, see region 1 in fig 3.

## 3.2. Curvature analysis

The smallness of the angle which is between the lines joining the current point on the boundary, to boundary points on either side a given distance away from

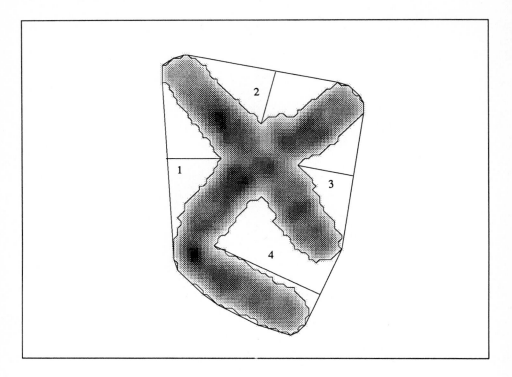

**Fig. 2.** The deepest or the most concave point is not necessarily a cut point.

it or the bigness of the curvature at the current point can be used as a measure of the "concavity" at the point [14]. Using the methods introduced by [1,3,8,10], a list of points at which the boundary forms a big curvature can be obtained. The four cut points are chosen using one of the following methods.

a) In four biggest concavities choose the most concave point (the angle at the point is smallest or the curvature at the point is biggest) as the cut point. The result of this method is nearly identical to the concavity analysis method and has the same problems. See region 4 in fig 2.

b) choose two candidates in each of four or five biggest concavities. Combine every set of four possible points which come from different concavities. The four points form a quadrilateral. The perimeter and area are then calculated. The set which either has minimum perimeter or minimum area is chosen. This method is better than those above, for example, the cut points of fig 2 can be found by this method.

c) Get the full list of concave points. Construct a set of 4 points using either the above criteria or the criterion that they approximate the vertices of a parallelogram, with minimum area.

The disadvantage of these methods is that there are number of counter-

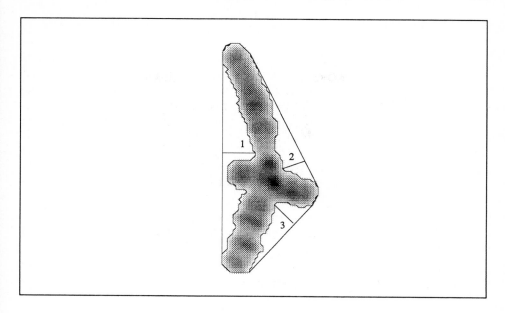

**Fig. 3.** There are *two* cut points in concavity 1.

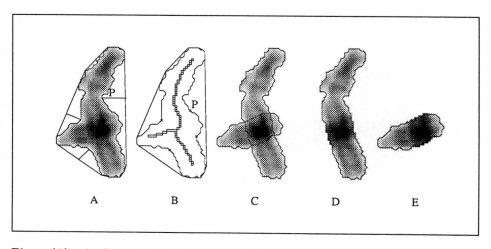

**Fig. 4.** (A) point P found by either method of deepest point or most concave point was not the correct cut point. (B) P was found correctly by using the skeleton node as a reference point. (C) separating curves found by the linear least squares method. (D,E) the resulting separated chromosomes.

examples and they do not always find the best candidates. For example, the cut point P in fig 4B is neither deepest nor "most concave".

### 3.3. Finding cut points using convex hull, skeleton and concavity analysis

Since the above two methods only use the information directly from the boundary and the criteria for choosing cut points are not very reliable, we decided that we should use new features to confirm or reject the cut points, which were located by the above method, or to help to find them. From the enormous literature on thinning algorithms, [2,13,21,...], it is well known that the skeleton preserves most of the information about a contour. We expect a pair of overlapping chromosomes to have a cross-like skeleton. The node of the skeleton should be inside the overlapping region. If we use this node as a reference point a more sensitive and accurate cut point finding procedure can be derived. The method depends on two hypotheses:

a) where chromosomes overlap, four cut points can be found.
b) where chromosomes overlap, there is at least one node in the skeleton of the object.

The goal of the procedure is to find the four cut points then use them to decompose the overlapping chromosomes.

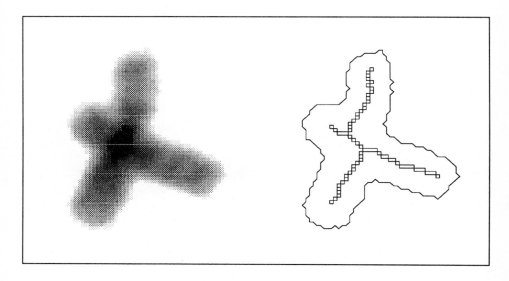

**Fig. 5.** The skeleton of this pair of overlapping chromosomes has two nodes

**3.3.1. Reference point finding.** We mentioned before that the node of the skeleton can be used as a reference point. There is a lot of literature giving methods to find skeletons [2,6,13,17,21], but the details are beyond the scope of this paper. We only suppose that the skeleton is obtained in some way. Usually the skeleton is very sensitive to the fluctuations of the boundary and the result is a lot of branches and nodes. We can get rid of the noise by either smoothing the boundary first or using a multi-resolution skeletonisation algorithm. And we have noticed that a very smooth overlapping object may have two nodes to represent the intersection ( see fig 5 ). So this section will give the method of finding a list of possible reference points. In next section we discuss the method of finding which one is the real reference point.

We can get a list of candidate reference points from the skeleton of the object. If the skeleton has no node, then the object is regarded as unsplittable by this technique. Otherwise the procedure, expressed in "pseudo-code", is

```
for each (skeleton node)
      for each (branch of the node)
            if (the branch is to an end point)
                  get the length of the branch;
            endif
      endfor
      choose the shortest one as "associated
          length" of the node;
endfor
sort the nodes according to their "associated
      length"; # longest first
return (the number of nodes found and their locations);
```

If more than one candidate appears in the list we leave which one is the real reference point undecided until it is confirmed by the next procedure. We test them in the order of the list.

**3.3.2. Cut point finding from reference point.** The next step is to find the four cut points. Because the reference point indicates where the overlapping region is, we only consider potential cut points which are "near" the reference point. We first find a set of possible cut points, and then select the "real" cut points from this set, so far as is possible choosing points which lie in different boundary concavities. In "pseudo-code" the procedure is

```
construct the convex hull of the object;
find five longest chords from the convex hull;
# The reference points are ordered such that "most central" is tried first
for each (candidate reference point)
      # Get list of potential cut points P_{ij} and associated
      # distances d_{ij} to the reference point
      for each (of the five chords) # indexed by j=1,2,3,4,5
            find the piece of boundary whose ends are those
                of the chord;
            calculate the distance from the reference point
                to the points on the piece of boundary;
```

```
            get two shortest local minima;
            put these two points Pij (i=1,2) and associated
                distances dij in the "cut point list";
        endfor
        # choose threshold for selecting cut points from list {Pij}
        thresh = k * min(dij);        # k is a suitable constant
        Mark the points which are nearer to the reference point
            than the threshold, put them into a ''mark-list'',
            and get the number of marked points m_num;
        # Find four cut points from the "mark-list"
        if (m_num ≥ 4)
            # Four or more cut points are marked.
            # Count how many concavities are involved.
            c_num = 0;        #number of cut point concavities
            for each (of the five long chords j)
                    if (either the corresponding P1j or
                        P2j is marked)
                            c_num = c_num + 1;
                    endif;
            endfor
            if (c_num ≥ 4)
                    # Either four or five concavities involved.
                    # From each concavity just one cut point is chosen.
                    choose four (or five) points from the "mark
                        list", each of which comes from a
                        different piece of the boundary associated
                        with one of the five chords;
                    # procedure exits returning four cut points
                    return (the four which have shortest
                        associated distance)
            elseif (c_num = 3)
                    # two cut points must be chosen from the
                    # same concavity
                    while (m_num > 4)
                            remove marked point with biggest Dij;
                            m_num = m_num - 1;
                    endwhile
                    # procedure exits returning four cut points
                    return (the remaining four cut points);
            endif
        endif
        # If unsuccessful, loop using the next reference point (if any)
    endfor
```

```
# Failure exit : four cut points cannot be found;
# the object is apparently not an overlapping cluster.
return (failed);
```

## 4. Decomposition

After the four cut points have been obtained, the rest of decomposition procedure becomes straightforward. We describe it briefly.

Firstly the missing parts of the boundary should be found, then they are used for separation of the overlapping chromosomes.

8 adjacent segment pieces (5 - 10 pixels) of boundary each side of the cut points can be found by following the boundary clockwise and anti-clockwise. Two pieces of boundary form a pair that are in a sense "opposite". The missing part of the individual chromosome's boundary, which is between these two pieces, can be obtained by linear least square method from the pair.

Fig 6 shows an example of finding missing part of boundary. Searching from the cut points $C_1$, $C_2$, $C_3$, $C_4$ along the boundary clockwise, segment pieces of boundary $b_{21}$, $b_{32}$, $b_{43}$, $b_{14}$ can be obtained; and similarly $b_{41}$, $b_{12}$, $b_{23}$, $b_{34}$ by searching anti-clockwise. See fig 6b. The missing parts of the boundary $(C_i \ C_j)$ can be obtained by linear least square method from the pair $b_{ij}$ and $b_{ji}$ (i, j = 1, 2, 3, 4; i $\neq$ j). Fig 6c shows the result from cubic curve fitting. Using the pieces of the boundary which are between a pair of adjacent cut points, the overlapping chromosomes can be separated. Fig 6a, 6d and 6e are the original cluster and the separated chromosomes.

## 5. Holes

If there is a hole in an object, there will be a loop in the skeleton. It makes finding the correct reference point more difficult. We do not here attempt to give details of how to deal with holes, but give some simple examples instead. The details can be found in [14].

If the hole is very small or very near the boundary, we can get rid of it by either "filling" the hole [14] or using dilation and erosion. If a "big" hole is inside the object we can "open" it to the external boundary [14]. Fig 7 shows an interesting example in which both touching and overlapping are involved. After the hole is "opened" the separation can be achieved by either splitting the cluster first then decomposing the overlapping one, or decomposition first then splitting the remaining cluster.

## 6. Equipment and data

The data comprised 46 overlapping G-banded chromosomes manually identified in 125 cells. A few overlapping orcein stained unbanded chromosomes selected

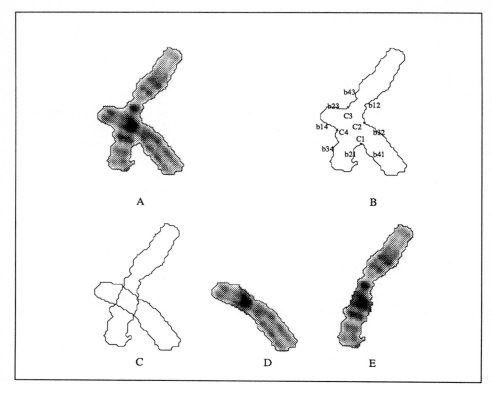

**Fig. 6.** (a) A pair of overlapping chromosomes. (b) the 8 labelled segments of the bound-ary are obtained by following the boundary clockwise and anti-clockwise from the four cut points C1, C2, C3, C4. (c) The missing pieces of the individual chromosomes' boundaries were obtained by fitting cubic curves using the linear least squares method from the pairs of boundary segments (b21,b12), (b32,b23), (b43,b34) and (b14,b41). (d,e) the resulting segmented chromosomes.

visually were also tested. The G-banded cells were obtained using a TV camera attached to a CRS-4000 frame store on a PDP11/34. The orcein stained cells were selected and digitised automatically on the MRC's FIP machine. The FIP stage permits fine scanning at about $0.1\mu$m pixel spacing [20]. The 8-bit density data was thresholded using a global threshold and segmented into connected components. The threshold level is determined using a smooth copy of the pixel intensity histogram, finding the first zero-slope point which is less than one third of the peak height. The above threshold objects were linearly normalised so that their combined pixel intensity histogram ranged from value 1 to 255. The splitting program was implemented on a VAX-780 and also on a M68000 microcomputer.

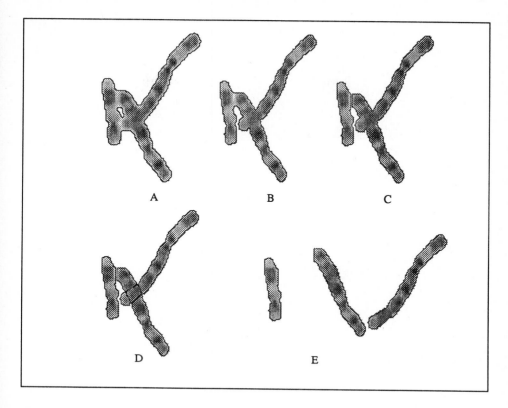

**Fig. 7.** (A) A cluster with a hole. (B) the hole is opened to the boundary. (C) the non-overlapped chromosome is split off [13]. (D) the overlap is decomposed. (E) the resulting separated chromosomes.

## 7. Results

The following table shows the result for the 46 G-banded pairs of overlapping chromosomes.

|          | objects submitted | correct split | wrong split | fail to split |
|----------|-------------------|---------------|-------------|---------------|
| Number   | 46                | 43.5          | 0.5*        | 2             |
| %        | 100               | 94.6          | 1.1         | 4.3           |

* One chromosome of an overlapping pair was separated correctly but the other was slightly wrong.

In the test data each of the overlapped clusters comprises only two chromosomes. Actually the procedure can be used in more complex cases. Fig 8 shows an example in which three chromosomes are overlapping. They still can be separated provide the procedure is used twice.

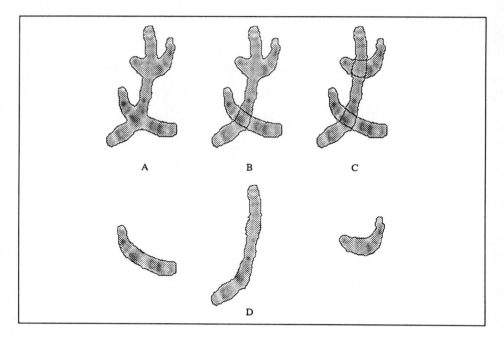

**Fig. 8.** (A) three overlapping chromosomes. (B,C) the first and second decompositions. (D) the resulting chromosomes.

## 8. Discussion

The high success rate shows that the procedure works very well for the G-banded overlapping chromosomes. Because the grey level has not been used in the procedure, it is obvious that there will be no difference when it is applied to other stained chromosomes. A few overlapping orcein stained unbanded chromosomes were also tested. The success rate is nearly the same as banded chromosomes but is not included in the table.

The reader will appreciate that the grey level of the original overlapping region will certainly be wrong in separated G-banded chromosomes. We can not simple distribute the grey value evenly or according the distance to the centre [12], because the chromosomes have a banding pattern. Before classification of the chromosomes we can not be sure what the pattern is, but the classifier can work only after the objects are segmented. A similar dilemma exists when two chromosomes overlap incompletely (fig 9). We do not know how deep one chromosome is inserted into the other. If the length of the inserted chromosomes is available we can then calculate the correct split. Unfortunately the length is also unknown until we know which class it belongs to. The best solution may be that rearrange the whole system, allowing segmentation and

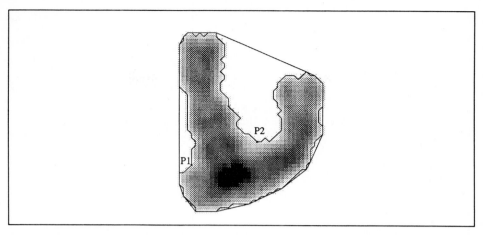

**Fig. 9.** Two chromosomes which are "incompletely overlapped". Only the two cut points P1, P2 can be found, and correct decomposition is therefore not possible.

classification at the same time with feedback between them. But that is beyond the limit of a short paper.

A very useful side production from the procedure is that it can be used as a means of auto-detection of overlapping chromosomes. Existence of a skeleton node, 4 cut points, and the mean density in the overlapping region (which is defined by the quadrilateral whose vertices are the four cut points) are certainly three powerful features to determine if an object is a cluster of overlapping chromosomes. We will show in a paper in preparation that a fully auto-detection and auto-segmentation procedure can be obtained by use of the method described here and the procedure in [14]. The success rate of the resulting segmentation is nearly as high as by a human operator.

*Acknowledgements.* I am grateful to Denis Rutovitz and Jim Piper for advice, comments and encouragement.

# References

1    Anderson IM, Bezdek JC. Curvature and Tangential Deflection of Discrete Arcs: A Theory Based on the Commutator of Scatter Matrix Pairs and Its Application to Vertex Detection in Planar Shape Date. IEEE Trans. PAMI-6:27-40 (1984)
2    Arcelli C, Di Baja GD. A width-independent Fast Thinning Algorithm. IEEE. Trans. PAMI-7:463-474 (1985)
3    Bowie JE, YoungIT. An Analysis Technique for Biological Shape - 2. Acta Cytologica 21:455-464 (1977)
4    Brenner JF, Necheles TF, Bonacossa IA, Fristensky R, Weintraub BA, Neurath PW. Scene Segmentation for the Analysis of Routine Bone Marrow Smears from Acute Lymphoblastic Leukeamia Patients. J. Histochem. Cytochem. 25:601-613 (1977)

5     Chassery JM, Gaybay C. An iterative Segmentation Method Based on A Contextual Color and Shape Criterion. IEEE Trans. PAMI-6:794-799 (1984)

6     Davies ER, Plummer APN. Thinning algorithms: A Critique and a New Methodology. Patt. Recognition 14:53-63 (1981)

7     Eccles MJ, McQueen MPC, Rosen D. Analysis of The Digitized Boundaries of Planar Objects. Patt. Recognition 9:31-41 (1977)

8     Freeman H. Shape Description via The use of Critical points, Patt. Recognition 10:159-166 (1978).

9     Gaybay C. Image Structure Representation and Processing: A Discussion of Some Segmentation Methods in Cytology. IEEE Trans. PAMI-8:140-146 (1986)

10    Gallus G, Neurath PW. Improved Computer Chromosome Analysis Incorporating Pre-processing and Boundary Analysis. Phys. Med. Biol. 15:435-455 (1970)

11    Graham J. Resolution of composites in interactive karyotyping. This volume (1989)

12    Hilditch J. An Application of Graph Theory in Pattern Recognition. In: Michie D (ed) Machine Intelligence 3 (Edinburgh University press, 1968) pp.325-347

13    Hilditch J. Linear skeleton from square cupboards. In: Meltzer B, Michie D (eds) Machine Intelligence 4 (Edinburgh University press, 1969) pp.403-420

14    Ji L. Intelligent Splitting in the Chromosome Domain. Patt. Recognition (in press, 1989).

15    Lester JM, Williams HA, Weintraub BA, Brenner JF. Two Graph Searching Techniques for Boundary Finding in White Blood Cell Images, Comput. Biol. Med. 8:293-308 (1979)

16    Ledley RS, Ruddle FH, Wilson JB, Belson M, Albarran J. The case of the touching and overlapping chromosomes. In: Cheng GC, Ledley RS, Pollock DK, Rosenfeld A(eds) Pictorial Pattern Recognition (Thompson, Washington DC, 1968) pp.87-97

17    Pavlidis T. A Thinning Algorithm for Discrete Binary Image. C.G.I.P. 13:142-157 (1980)

18    Pavlidis T. Algorithms for Graphics and Image Processing. (Springer-Verlag, Berlin, Heidelberg, New York, 1982)

19    Piper J, Lundsteen C. Human chromosome analysis by machine. Trends in Genetics 3:309-313 (1987)

20    Shippey G, Carothers AD, Gordon J. The Operation and Performance of An Automatic Metaphase Finder Based on The MRC Fast Interval Processor. J. Histochem. Cytochem. 34:1245-1252 (1986)

21    Smith RW. Computer Processing of Line Images: A Survey. Patt. Recognition 20:7-15 (1987)

22    Vanderheydt L, Dom F, Oosterlinck A, Van Den Berghe H. Two dimensional shape decomposition using fuzzy subset theory applied to automatic chromosome analysis. Patt. Recognition 13:147-157 (1981)

23    Vossepoel AM. Separation of touching chromosomes. This volume (1989)

# Resolution of Composites in Interactive Karyotyping

James Graham

## Summary

The major requirement for operator interaction in automated karyotyping systems arises from the formation of "composite chromosomes" at the segmentation stage. These composites occur because chromosomes often touch closely or overlap so that they cannot be separated by simple, context free, segmentation methods such as thresholding, however carefully applied. This paper describes a method for automatically resolving the most commonly occurring type of composite, that due to touching chromosomes, by using simple contextual information about chromosome positions to drive a split and merge algorithm. A pilot trial shows that application of this method results in a significant reduction in the need for interaction in the analysis of metaphase cells of routine quality.

## 1. Introduction

Recently there has been an increase in interest in interactive karyotyping as a useful technology in the routine clinical chromosome laboratory. The system which I have described at previous workshops of the European Concerted Action in automated cytogenetics [1,2] has been available as a commercial product for some time. Other products with a similar functional specification are available from a number of manufacturers, both in Europe and the United States. Some of these products are packages of integrated metaphase finders and karyotypers. Some are karyotyping only systems. Lundsteen and Martin [3] have recently produced a useful review of the systems available and their various strengths and weaknesses.

Interactive karyotyping is the process of machine analysis of high resolution images of metaphase or pro-metaphase cells (usually from blood or amniotic fluid), supervised by a trained cytogeneticist or cytotechnician. The analysis consists of segmentation of individual chromosomes, measurements of size, banding pattern and centromere position, classification into pairs and, usually, production of a karyogram which can be reproduced on some hard copy device. At each stage in the analysis fairly standard pattern recognition techniques provide reasonably satisfactory results. However, in a system required to deal with the range of sample qualities obtainable routinely, and the occurrences of overlaps, twisted chromosomes, debris etc. that occur in eventhe best prepared

**Automation of Cytogenetics**   Editors: C. Lundsteen J. Piper
© Springer-Verlag Berlin Heidelberg New York 1989

specimen, reasonably satisfactory results still leave a great deal of scope for error. Misclassifications occur, centromeres are incorrectly located and, most importantly for the purposes of this paper, chromosomes can be poorly segmented. All of these difficulties are resolved by allowing the skilled operator to intervene in the analysis and correct errors. From one point of view this is a very acceptable solution. It allows the skills of the operator to be employed efficiently in the analysis, but it requires that the interactions are straightforward and meaningful to an operator who has a deep understanding of the image, though not necessarily of the machine, and it also requires that the number of interactions needed should not be excessive.

Interactive karyotyping software may be implemented on moderately priced general purpose hardware, allowing a useful measure of automation to be developed and widely used in clinical laboratories which do not have large capital equipment budgets. The system, which is the subject of this paper [4] was developed, together with a metaphase finder [5] on a Magiscan image analyser from Joyce Loebl. The hardware architecture of this instrument has been fully described by Taylor et al [6] and its data structures and image processing software by Graham et al [7]. This integrated hardware and software architecture makes for straightforward and efficient implementation of the type of algorithm described here.

In a clinical assessment of this system, Philip and Lundsteen [8] found that interaction time is the most significant part of the analysis and that operator happiness decreases rapidly with the number of interactions. By far the largest number of interactions occurs in the separation of composites of touching chromosomes, exceeding, for example, those needed to separate overlaps or to correct centromere positions.

This paper addresses the problem of reducing the number of interactions required in the segmentation phase by identifying and resolving composites which occur where several touching chromosomes are segmented as one object.

## 2. The Image Segmentation Process

For the purposes of this paper it is necessary to describe the segmentation procedure and its limitations in some detail. The images consist of $512 \times 512$ pixels of 64 grey levels obtained by a TV camera mounted on a microscope. There are two phases to the segmentation - a "coarse" segmentation in which the positions of objects are determined and the chromosomes are counted, and a fine segmentation where boundaries are located as accurately as possible, forming the first stage in the detailed analysis of the cell. Both segmentation phases are achieved by binary slicing at thresholds obtained from grey level histograms.

At the coarse "counting" segmentation, the histogram is taken from the central quarter of the image. A global threshold is selected in order to under-detect objects. The exact boundary positions are not important at this stage;

it is more important to try to ensure that closely touching chromosomes are located separately. The result is an approximation to the chromosome count; an approximation, because in most cells not all touching chromosomes will have been separated by this simple procedure, and some single chromosomes may have been split along light bands. These errors are corrected by an interaction. The automatically counted objects are each marked with a dot and the count is displayed. The operator corrects the count by indicating undetected or incorrectly detected objects with a lightpen, causing either the removal of the object or the marking of a new object, with a corresponding adjustment of the displayed count. This method of counting is efficient in that it makes use of the strengths of both the machine and the operator. Marking an internal point in each chromosome not only provides visual feedback to the operator during interaction, but also provides useful data for the next segmentation phase.

Fine segmentation locates the boundary of each chromosome by using a threshold obtained from a histogram of the image in the region close to it. The localities of the regions to be histogrammed are determined from the chromosome positions found at the coarse segmentation phase. The boundaries obtained by local thresholding are usually very satisfactory from the point of view of separation of chromosomes from background. However, the region of background between closely abutting chromosomes is often darker than the best threshold, resulting in composite chromosomes at the fine segmentation stage (figure 1a). Notice that this is not just a thresholding problem. A threshold which will separate touching chromosomes will almost certainly not result in a good overall boundary (figure 1b). The interaction required to resolve such composites is straightforward; the object is selected by pointing at it with the lightpen and the correct boundary positions indicated by drawing a single line. These steps, and the rest of the automated karyotyping process, are described in detail elsewhere [4].

## 3. Resolving Composites

In the case of closely abutting chromosomes, there is usually a valley in the grey value landscape between the two significant objects. The composite is created because the grey value in the valley floor is higher than the selected threshold. The position of best separation of the chromosomes is at the grey level minimum along the valley floor. Finding this minimum is functionally identical to the process of finding "watersheds" in grey level landscapes as described by the proponents of mathematical morphology [9]. Many morphological operators are cellular logic operators, and most efficient if implemented on a machine where a number of pixels are processed in parallel. Friedlander and Meyer [10] have described an algorithm for finding watersheds by a propagation method, suitable for pixel-serial processing. This algorithm involves first finding positions of grey level minima as "seeds" and propagating the regions of these

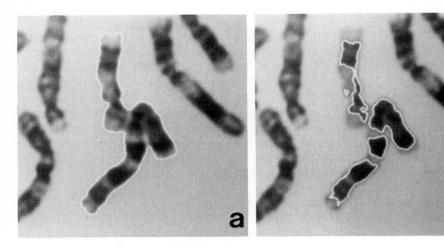

**Fig. 1.** A composite of two touching chromosomes. (a) The boundary obtained by local thresholding. (b) The boundaries obtained at a threshold which just separates the two chromosomes. Finding good boundaries for both chromosomes is not merely a thresholding problem.

seeds towards the watersheds. The second part of this algorithm is identical to a procedure described by Rutovitz in 1978 [11] for obtaining what he called (using the opposite topographical metaphor) the "fall-set" of some seed region; the fall-set is the set of points which can be reached by a downhill path from the seed region. The fall-sets of two neighbouring objects meet and terminate at the minimum of the grey level valley between them .

In the case of chromosomes, we wish to be careful about which minima we find, since the chromosome bands also give rise to grey level valleys, where we would certainly rather not split an object. Suitable seed regions are the regions obtained by binary slicing the image at a threshold which just separates the two neighbouring chromosomes (figure 1b). In many cases this will mean that internal minima will lie within the seed regions and will not be subjected to the fall-set finding procedures. There is, however, the problem of deciding which objects require to be "just separated" and how to obtain a threshold which will achieve this. This is described in the "splitting" procedure below.

It may not always be possible to keep internal minima within the seed regions. Sometimes the minima between dark chromosome bands are lighter than the minima separating objects (figure 1b). In these cases splitting the chromosome is unavoidable, resulting in neighbouring fall-sets being fragments of the same chromosome. These fragments have to be re-merged, as described in the "merging" procedure below. Both splitting and merging use information about expected chromosome positions obtained at the counting stage. These processes are illustrated by the resolution of the "cartoon" composite in figures 2 and 3.

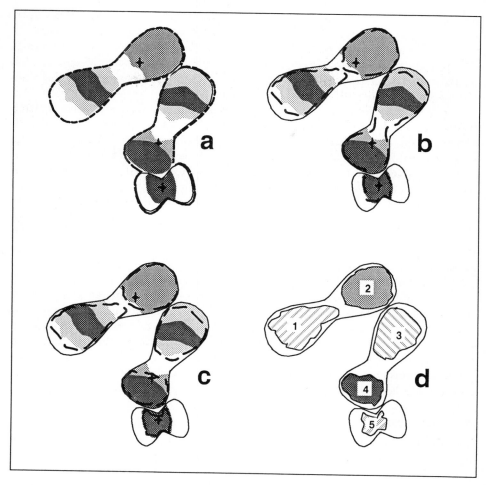

**Fig. 2.** Illustrating the splitting process. (a) The best threshold produces a boundary (dashed line) which encloses three chromosomes. The positions marked with a dot are those determined at the counting stage. (b) At a higher threshold the object starts to split up. The first split does not produce any new "core" fragments. (c) After several successive rethresholding passes, the object is split into five fragments, three of which are "core" fragments. (d) The "seed" regions obtained by the recursive splitting process, labeled for re-expansion.

## 3.1. Splitting

The boundary obtained by thresholding using a local histogram is taken as the best separation of chromosome from background. The data structures for describing image regions [7] allow this boundary and the region contained within it to be stored in such a way that image operations can be performed with maximum efficiency on only those points which form part of the boundary or region. One thing which can be easily checked is whether an identifying point obtained at the count phase is contained within the boundary. If, for example, there are no such points, then the object was not included in the original count and can be rejected as a potential chromosome. If a single point is found inside the boundary, then the object detected is a single chromosome and can be placed directly on the chromosome list. If more than one point is found then the object must be split by re-thresholding.

The re-thresholding can be thought of as proceeding recursively. At each stage the threshold is increased by one grey level, and a new binary slice is obtained only from those points inside the object obtained at the previous threshold value. If an object is split by re-thresholding, each new component is re-thresholded in turn. The splitting stops for a particular object when the number of count-points it contains is either one or zero. Whether this process ends in a single chromosome being isolated or in the creation of several fragments depends on the relative grey levels between light bands and the inter-chromosome valleys and the positions of the count-points (figure 2). This splitting process can be described by a tree structure where the nodes represent splits and the leaves represent the final seed regions. The threshold at which a particular split occurs is recorded at the appropriate node in the split tree.

Each of the regions extracted by recursive re-thresholding acts as a "seed" for expansion to its fall-set using Rutovitz's algorithm, within the boundary of the original composite object. The algorithm begins by writing a label value for each seed region into a label image (figure 2d), and at the end this label value has been propagated over the entire fall-set (figure 3). The fall-set of a particular seed region can be obtained by binary slicing the label image at the appropriate label value.

## 3.2. Merging

The fragments of the original composite arising from re-thresholding and expanding can be divided into two types: "core" fragments, i.e. those which

---

Fig. 3 (opposite). Illustrating the merging process. (a) Regions 1 - 5 are the results of expanding the "seeds" obtained in figure 2d using the fall-set algorithm. Not only do region boundaries occur between individual chromosomes, but two chromosomes have been split into abutting fragments. (b) The region adjacency matrix corresponding to this set of regions. The numbers indicate the likelihood of merging computed from the length of shared boundary and the image intensity at that boundary. "Core" fragments are marked with an asterisk. (c) The most likely merge (1 + 2) has taken place. Fragment 2 gains no new neighbours from this merge. The likelihood of a merge between 2 and 3 has gone down due to the decreased length of the common boundary as a fraction of the total boundary length. (d) The next most likely merge would involve two "core" fragments and is disallowed. Of the two possible merges involving fragment 3, the most likely is 3 + 4. (e) Fragments 3 and 4 are merged. Fragment 4 now gains fragment 2 as a neighbour. The likelihood of merging 4 and 5 is decreased due to the change in boundary length. The merging stops here as only "core" fragments remain.

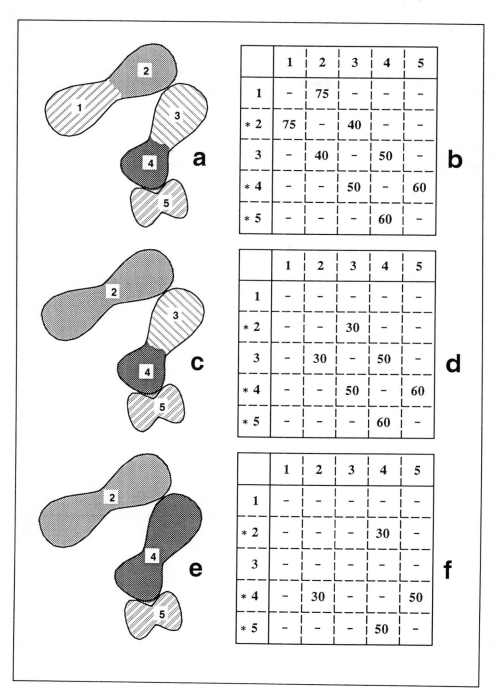

ended up containing a single count-point, and "non-core" fragments which ended up containing no count points. The core fragments may represent whole chromosomes or parts of a chromosome; the non-core fragments certainly represent parts of chromosomes and need to be re-merged with neighbouring core fragments to produce complete chromosomes. For each region it is possible to obtain from the label image the labels of neighbouring regions and the lengths of the common boundaries. From the splitting tree it is possible to determine at which threshold value that region separated from each of its neighbours. Both of these values are used in determining a likelihood for remerging each pair of regions.

A region adjacency matrix R is constructed in which each element $R_{ij}$ is a record containing the common boundary length, separating grey level, and merge likelihood for regions $i$ and $j$. As merging proceeds all of these values get updated. R is, of course, symmetric with $R_{ij} = 0$ for $i = j$. Each row is labeled as either a core or non-core fragment and adjacency values between neighbouring core fragments are initialised to zero to avoid re-merging two separated chromosomes. The matrix R not only indicates directly which merges are available, but can also be easily updated when a merge has occurred. If an object $i$ is merged with an object $j$ then the elements of row $j$ are moved into row $i$ so that previous neighbours of $j$ become neighbours of $i$. If elements $R_{ik}$ and $R_{jk}$ are both occupied, this means that object $k$ is a neighbour of both $i$ and $j$. The new $R_{ik}$ is updated by adding together the boundary lengths and substituting the minimum grey level separation in $R_{ik}$ and $R_{jk}$. The merge likelihoods for row $i$ are recalculated, row $j$ has all adjacencies set to zero and the matrix is made symmetric again. The merge is recorded in the label image by relabeling object $j$ as $i$ (figure 3).

Merges proceed in order of likelihood. The most likely merges occur first since the adjacency values will then have to be updated, affecting the likelihood of other merges. Quite often a non-core region will only have a single core neighbour and merging can proceed without difficulty. In other cases a choice must be made between potential merges. The most likely merge is chosen, of course, but only if the difference in likelihoods exceeds some threshold; the local criteria of grey level and boundary length are not very discriminating and only clear choices are made.

Merging continues in this way until either all non-core fragments have been united with neighbouring cores or the remaining non-core fragments have approximately equal likelihoods of merging with two or more neighbours. The merge criteria do not allow an informed decision to be made in these cases and a safe solution is adopted which returns the difficult case to the operator to be resolved interactively. This is achieved by a second merge pass in which the significance criterion is abandoned and neighbouring core regions are allowed to re-merge. Thus, core fragments which abut via these undecided non-core regions are forced to unite to form composites. Other core fragments not involved in these adjacency relationships remain separate, so that good work is not all undone. Figure 4 shows the result of applying this procedure to the composite object shown in figure 1.

### 3.3. Difficulties

Certain situations may occur which do not result in correct segmentations. Overlapping chromosomes, for example, do not fit the initial premise that the objects to be identified are separated by a grey level minimum. Splitting and merging will be attempted however, and, invariably, fail. The best outcome is a complete re-merge to come out with the original object again, but it is possible for the object to be split up into regions, none of which corresponds to a single chromosome and some of which may incorporate parts of more than one chromosome. Incorrect choices in re-merging can occasionally produce similar results in touching chromosomes . A special form of interaction is included in the karyotyping program to enable efficient resolution of this problem at the interaction stage (see below).

Very large composites can result in a large and complex adjacency matrix if the splitting proceeds to completion. The likelihood of erroneous remerging increases as the matrix becomes more complex, and the additional time required to perform the remerging is partially wasted. It is useful therefore to be able to stop the splitting at some controllable level of complexity. This is difficult to achieve if the splitting is truly recursive, as implied above, since recursive procedures are necessarily depth-first. An iterative, breadth first, algorithm is used here : all the nodes at a given level in the splitting tree are expanded simultaneously. The splitting can be stopped at any required level of complexity so that large composites are typically not resolved completely, but split into smaller composites. Useful side effects are that the iterative procedure is faster and the use of memory resources can be controlled.

## 4. A pilot trial

The results to be expected from the inclusion of this split and merge procedure in the karyotype analysis program are a reduction in the number of interactions required to achieve a correct segmentation, a reduction in the interaction time required and an increase in the time spent in automatic segmentation. The net result should be an overall decrease in the time taken to perform a complete analysis. To test the effectiveness of the method, fifty cells (25 blood and 25 amniotic fluid) were selected from lists provided by the metaphase finding component of the same package [5]. Some selection was applied to avoid analysing the one or two very poor quality cells presented in the list, but the twenty five cells used came from the first twenty nine as selected by the metaphase finder in the amniotic case, and the first twenty six in the case of blood and so can be claimed to be representative of those which might be analysed routinely.

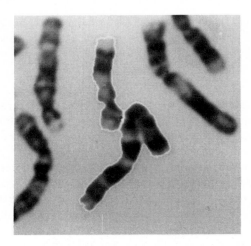

**Fig. 4.** The result of the split and merge process applied to the composite object in figure 1.

(Indeed in their clinical trial, Lundsteen et al [12] use 10 cells from the first 19 on average, only four of these being fully analysed. This is a much stricter testing environment.)

The cells were first analysed by a "thresholding only" version of the software and then by a "split and merge" version. Automatic segmentation time and interaction time were noted ,as well as the total number of composites for which interaction was required. The measurements applied only to the segmentation stage of the karyotype. Also noted were the number of overlaps which occurred, and which would not be expected to be resolved, and the number of cases of an incorrect merge requiring the use of a new "merge and redivide" interaction mode, not required in the "thresholding only" version. It is important to know that this new facility is not a net creator of interactions! The results are shown in table 1.

Two important figures determining the ease of interaction are the number of composites per cell and the number of chromosomes forming the composite (clump size). Small composites of up to three or four chromosomes may be separated with a single interaction The table shows that the number of composites per cell has been reduced by over 70%, the reduction being rather greater in the case of blood than amniotic fluid. The mean clump size has been reduced slightly, mainly due to the breaking up of very large composites. In fact only two composites consisting of ten or more chromosomes occurred with the split and merge software whereas thresholding alone resulted in ten of these large objects. Segmentation using thresholding alone resulted in the requirement for at least one interaction in all cells, and two or more interactions in all but two. Using the split and merge software, no interactions were required in 28% of cells and only one interaction in a further 18%.

**Table 1.** Summarised data on segmentation performance with and without the application of the split and merge procedures

| | AMNIOTIC | | BLOOD | | ALL | |
|---|---|---|---|---|---|---|
| | Thres-hold only | Split and merge | Thres-hold only | Split and merge | Thresh-hold only | Split and merge |
| Total number of cells | 25 | 25 | 25 | 25 | 50 | 50 |
| Total composites | 193 | 59 | 171 | 50 | 364 | 109 |
| Average number of composites per cell | 7.78 | 2.36 | 6.84 | 2.0 | 7.28 | 2.18 |
| Average clump size | 3.6 | 3.2 | 2.9 | 2.4 | 3.25 | 2.81 |
| Number of cells with no interactions | 0 | 5 | 0 | 9 | 0 | 14 |
| Number of cells with one interaction | 0 | 5 | 1 | 4 | 1 | 9 |
| Number of composites which include overlaps | 3 | 7 | 11 | 16 | 14 | 23 |
| Number of "merge and redivide" interactions | 0 | 6 | 0 | 11 | 0 | 17 |
| Average automatic time | 15.7 | 29.9 | 17.2 | 31.1 | 16.4 | 30.5 |
| Average interaction time | 85.5 | 40.7 | 59.4 | 34.9 | 72.5 | 37.8 |
| Average segmentation time | 101.2 | 70.6 | 76.6 | 66.0 | 88.8 | 68.3 |

The number of composites which included a chromosome overlap increased using the split and merge software. At first sight this may seem rather curious since the same cells were used in both tests. The explanation is that composites containing two or more overlaps have largely been broken up into composites containing only one. These comprise about 20% of the residual requirements for interaction. Of the seventeen cases in which the new "merge and redivide" interaction was required, six were due to the existence of overlaps. The remainder were due to incorrect decisions in touching chromosomes.

The total segmentation time has been reduced by some 20%-30%. Interestingly, the greatest time saving occurs in the case of amniotic fluid samples, where the remaining interactions are greatest in number. This must be attributed to the removal of the very large composites which require long interaction times.

## 5. Discussion

Karyotype analysis of metaphase cells has for a long time been a target for the application of automatic image interpretation methods. In recent years this effort has borne fruit in the form of a number of commercially available karyotyping systems, some with associated metaphase finders and some without. All of these systems require operator interaction to achieve complete analysis of all but the most perfect cells. Clinical trials on the system implemented on the Joyce Loebl Magiscan image analyser conclude that the main requirement for interactions is in the separation of touching chromosomes [8].

I have described here a method which automatically resolves a large proportion of the most commonly occurring cases of composite chromosomes and which can eliminate the need for interaction at the segmentation stage in significant proportion of cells of routine quality. The method has been aimed at resolving the composites which occur most frequently in blood and amniotic fluid samples, for which the analysis package was primarily intended, namely small clumps of touching chromosomes. The resolution of overlapping chromosomes almost certainly requires some syntactic shape analysis (see for example Ji [13] this volume) and may be the next most pressing target for reducing the number of interactions. Large agglomerations of touching chromosomes may be only partially resolved.

The results of a small pilot trial show that significant reductions in the requirement for interaction have been achieved and that this results in some time-saving in the overall analysis. The main advantage seems to be that the worst types of interaction for the operator, namely splitting large composites, have been effectively removed and that, in a substantial proportion of cells analysed, no interaction at all is required for segmentation. Lundsteen et al [12] have included a version of the software including the "split and merge" procedure in a test of clinical performance involving a large set of cells from the routine laboratory production. Several different software configurations were under test in this trial, and it is not easy to identify the improvement in interaction requirement due to segmentation alone. However, one can estimate that about 40% fewer interactions were required at the segmentation phase in their study. This improvement is somewhat less than would be expected from the pilot trial presented here. The difference may be due to a difference in the quality of preparation used in the two trials but this seems unlikely since the slides used here were obtained from the Rigshospital chromosome laboratory. The regime in the Rigshospital trial was to fully analyse four cells from the first nineteen detected and ranked by the metaphase finder. Thus the comparison consisted of analysis of relatively "easy" metaphases. The figures presented here suggest that this method produced a substantial improvement in the analysis of more difficult cells. One result of using this procedure may be a reduction in the number of cells which have to be located in order to perform the required number of full analyses.

The clinical trial did not produce figures on the number of cases in which no interaction was necessary at segmentation. The authors do however speculate on the possibility of using this version of the software as the basis of a "semi-hands-off" karyotyping system in which the operator only interacts at the count stage and to correct and verify the final karyogram.

The knowledge which drives this method is very straightforward - a list of points, each one internal to a single chromosome. This knowledge is available in the system for other purposes and its use here has been rather opportunistic. The simplicity of this knowledge is responsible for some of the residual errors. The re-merging process for instance would be much more reliable if the objects created by merging where tested for their plausibility as chromosomes, rather than simply using local context-free information about the shared boundary between merged fragments. There is some interest now in the use of "knowledge based" methods derived from the field of artificial intelligence in generating a truly automatic karyotyping system. Piper et al [14] give a very clear account of the complex set of relationships between objects and the knowledge about those objects which arise in the karyotyping problem. A complete specification of this knowledge and these relationships seems a daunting project. This study indicates that the application of explicit knowledge locally to specific tasks such as composite resolution has considerable promise for success.

# References

1.   Graham J. The Magiscan Interactive Chromosome Karyotyper - MICKY. Proc. V European Chromosome Analysis Workshop, Heidelberg (1983)
2.   Graham J. Recent Developments on the Magiscan Karyotyping System. Proc. VI European Chromosome Analysis Workshop, Leiden (1985)
3.   Lundsteen C, Martin AO. On the selection of Systems for Automated Cytogenetic Analysis. Am. J. Med. Genet. In press.
4.   Graham J. Automation of Routine Clinical Chromosome Analysis I. Karyotyping by Machine. Anal. Quant. Cytol. Histol. 9:383-390 (1987)
5.   Graham J, Pycock D. Automation of Routine Clinical Chromosome Analysis II. Metaphase Finding. Anal. Quant. Cytol. Histol. 9:391-397 (1987)
6.   Taylor CJ, Dixon RN, Gregory PJ, Graham J. An Architecture for integrating Symbolic and Numerical Image Processing. In: Duff MJB (ed) Intermediate Level Image Processing (Academic Press, London, 1986) pp.19-34
7.   Graham J, Taylor CJ, Cooper DH, Dixon RN. A Compact set of Image Processing Primitives and their Role in a successful Application Program. Patt. Recog. Lett. 4:325-333 (1986)
8.   Philip J, Lundsteen C. Semi-automated Chromosome Analysis. A Clinical Test. Clinical Genetics 27:140-146 (1985)
9.   Serra J. Image Analysis and Mathematical Morphology. (Academic Press, London, 1982) pp.456-463
10.  Friedlander F, Meyer F. A sequential Algorithm for Detecting Watersheds in a Grey Level Image. Proc. 7th Congress of ISS, Caen, France (1987)
11.  Rutovitz D. Expanding Picture Components to Natural Density Boundaries by Propagation Methods. The Notions of Fall Set and Fall Distance. Proc. IV IJCPR, Kyoto, Japan, (1978) pp.657-664
12.  Lundsteen C, Gerdes T, Maahr J, Philip J. Clinical Performance of a System for Semi-Automated Chromosome Analysis Am. J. Hum. Genet. 41:493-502 (1987)
13.  Ji L. Automatic Resolution of Overlapping Chromosomes. This volume.
14.  Piper J, Baldock R, Towers S, Rutovitz D. Towards a Knowledge Based Chromosome Analysis System. This volume.

# Separation of Touching Chromosomes

Albert M. Vossepoel

## Summary

An algorithm is presented for separating touching chromosomes, which is often necessary in bone-marrow metaphase analysis. The algorithm is intended as a completely automatic preprocessing step preceding further interaction with the metaphase image. It is primarily based on shape analysis of the object contours found after conventional thresholding of the image. Inward bends in the contours are considered as potential starting and target points for cutting lines. The algorithm being sequential, the order in which objects are cut into chromosomes is very important. The order is determined by the successive application of shape criteria. An evaluation scheme is also presented, providing guidance to the selection and ordering of these criteria.

## 1. Introduction

The analysis of chromosomal aberrations in bone-marrow slides is of great importance for the diagnosis of leukaemia. The character and frequency of the aberrations, such as aneuploidy or translocation, are decisive in the diagnosis and classification of the disease into main categories, each of which frequently require different treatment.

However, the analysis of bone-marrow chromosomes is generally hampered by the low frequency of metaphases in the samples, as a result of both the leukaemia itself and of the treatment with cytostatic agents, which in some cases has already started at the time of taking the bone-marrow sample. Culturing does not substantially improve this. Not only are fewer metaphases found on a slide, but generally these metaphases are also of a much poorer quality than those from blood cultures. The chromosomes tend to be shorter and less well-spread, often lying alongside or on top of each other, and there is sometimes other material present in the close vicinity, such as nuclei and debris.

On the basis of the number of leukemia investigations we estimate that in the Netherlands about 4000 analyses of bone-marrow chromosomes will be required yearly in the near future, each analysis taking several hours of tedious labour for a well-trained cytotechnician. It is evident that some computer assistance would be welcome, in the first place for finding metaphases, and later on in karyotyping every metaphase found, as far as possible.

**Automation of Cytogenetics**   Editors: C. Lundsteen J. Piper
© Springer-Verlag Berlin Heidelberg New York 1989

In recent years automated metaphase finding in slides from blood cultures has much improved. It is probable that this can also be developed for bone-marrow preparations, since the problems are similar, but have a greater degree of complexity for the reasons given above. Here we will focus on the karyotyping process. In spite of the great expectations of the scientific community some 20 years ago, this process still has not become fully automated for slides from blood cultures, and certainly not for those from bone-marrow samples.

As every piece of information from the few available metaphases in bone-marrow preparations is needed, one cannot afford to discard poor ones. This requires putting more emphasis on a proper separation of the individual chromosomes, even if they are very close to, or even on top of each other. When performed manually, be it with a pair of scissors and a photograph, or with an electronic device such as a light-pen on a television screen, this is a tedious and time-consuming task. Therefore, the automation presented here aims at separating chromosomes with as little interaction as possible.

## 2. Materials and Methods

### 2.1. Hardware

The Genetics Department at Leiden uses a Magiscan-2 provided by the manufacturer, Joyce-Loebl Ltd. The equipment is attractive not only because its architecture allows fast image processing, but also because it operates under the UCSD Pascal System, making it accessible for programming in Pascal, and the availability of various chromosome analysis algorithms developed by Graham [1,2,3] as a starting point. The system consists of a processor of proprietary design with bit-sliced architecture, but without an interrupt facility. The processor can manipulate image data in slices of variable size, up to $1024 \times 1024$ elements with a "depth" range from 1 to 8 bits. The slices can be defined by the user as sets of contiguous bit planes in a memory in which 8 of these planes, each with $1024 \times 1024$ elements, are available. A second memory of equal size can be fitted into the system. The contents of the image memory can be displayed through a window of $512 \times 512$ elements. The limitation in spatial resolution stems from the application of standard television technology, with 625 scan lines per frame, 50Hz interlaced, i.e., 50 half frames per second. The overall time available for one scan line is $64\mu s$, of which $11\mu s$ are required for trace-back. With an aspect ratio of 4:3 for the display screen, in the remaining $53\mu s$ a scan line, equivalent to $(4/3) \times 625$ picture elements has to be sampled, requiring a sampling frequency of 16MHz. The grey-value resolution of the image is 6 bits, on input as well as on output, limited by the capacity of the analog-to-digital (A/D) and digital-to-analog (D/A) converter, respectively. Both on input and on output, the grey-values are translated by a user-definable look-up table.

The input device is a Bosch TYK 9A chalnicon television camera, mounted on a Leitz Orthoplan microscope through a variable magnification ("zoom") attachment. The microscope is equipped with a Zeiss (Märtzhaüser) stage, on which 4 slides can be mounted. Except for changing objectives (and applying immersion oil) all functions of the microscope are controlled by the Magiscan-2 system. These functions are:

- Position control, by stage motion in $x$- and $y$-direction, with ranges of 108 and 54mm, in about 20900 and 10200 increments of $5.15\mu$m;
- Focus control, by stage motion in $z$-direction, with a range of 1 mm, in 10000 0.1$\mu$m increments;
- Illumination control, through the lamp power supply, in 100 increments;
- Magnification control, through the "zoom" attachment, with a range of $\times$1 to $\times$3.2 in 1500 increments.

The field of view of the television camera is restricted to less than half that of the the 20mm diameter microscope ocular: a square with sides of 8mm, both sizes to be divided by the magnification of the objective used, and by the "zoom" factor.

## 2.2 Primary segmentation software

Experience has taught that thresholding techniques can result in simple, and thereby robust and fast implementations on general-purpose hardware [4]. The time requirement is important because in the present routine the chromosome analysis still requires operator interaction at the metaphase relocation step. This interaction consists in adjusting the position, changing magnification, e.g., from $\times$16 to $\times$100, adjusting the "zoom" and the focus. Less interaction would be required if the metaphases found automatically were first all reviewed, selected and re-ordered at low magnification. Next, all selected metaphases can — in principle — be analysed automatically, provided the adjustments of position, magnification ("zoom") and focus can also be handled automatically. However, in that case the robustness is of paramount importance.

Even in simple, global thresholding, it is very difficult to select a threshold level that simultaneously minimizes fragmentation as well as clustering of chromosomes. Although thresholding cannot provide a perfect primary segmentation between objects and background, a well-chosen threshold can save a lot of trouble in a later splitting or merging stage for object segments. A reasonable choice is the grey-level at the inter-mode minimum of the histogram. Next, this level is "tuned" by applying a small global offset $\delta T$ in increments of one unit grey-level at a time.

In practice, most metaphase images require adaptive thresholding. This can be accomplished by delineating each object at a preliminary global threshold, e.g., half-way in between the object and background peaks in the grey-value histogram. Next, in the Magiscan-2 chromosome system, each object is placed in a rectangular frame with at least 3 times the object area. A local threshold is computed for each distinct frame. If, on the one hand, the local threshold is "darker" than the global one, it is only applied within the object. On the other hand, if the local threshold is "lighter" than the global one, it is also applied within the object, but now after this has first been dilated with one or more rims of pixels. Especially in the latter case, iteration of the local thresholding procedure — not provided by the present version of the Magiscan-2 software — may be necessary. For practical reasons, the existing adaptive thresholding has been replaced by an adaptive version of the global one. The image is subdivided into 8 × 8 rectangles, for each of which a threshold is computed. The local threshold is then found by — e.g. bilinear — interpolation between the nearest rectangle thresholds.

## 2.3. Valley Searching software

In the original segmentation software of the Magiscan-2, which has been devised for amniotic fluid samples instead of bone-marrow ones, the concept of "fall-set" [5] is employed in the object segmentation. In practice, the algorithm employed results in a fragmentation of the clustered chromosomes. The fragments then have to be reassembled in a following split-and-merge step, which is apparently intended to function under interactive guidance, i.e., it is presumed that the number of chromosomes in a cluster is indicated interactively. In order to suppress this interaction, and because the results of this software turned out to be rather disappointing on bone-marrow metaphases, a different approach has been devised.

Once objects have been formed using a "conservative" threshold, one can often detect from the shape of the object that it contains more than one chromosome. Then characteristic rather sharp inward bends are present in the contour, for which there exist excellent detection algorithms [6].

If, at such an inward bend, there is also a "valley" of relatively light grey-values running into the object, then the object can be cut by simply following this valley. Following becomes easier if the valley is shorter, or if two valleys running from different points of the contour form a "pass". Assuming that such a "pass" will occur in most cases, it makes sense to search in the direction of another starting point candidate on the opposite side of the contour, if available as a target.

In a first implementation of the valley-search a minimum-cost algorithm without targeting has been used, i.e., running like a river, which – thanks to Pascal – could flow recursively. The resulting valleys were, however, rather unpredictable and indeed meandered like rivers. The meandering could effectively be suppressed by limiting the progress of the valley search to only three of the eight elementary directions. If a target was available, the central direction of these three pointed approximately towards this target. But even then, many cutting lines missed their target.

In the present implementation, for every cutting line an origin and a target are chosen first. A candidate for origin or target is a local (concave) maximum — exceeding a distinct threshold $c_{min}$ — in the contour curvature. This quantity is obtained by convolving the contour's differential chain codes by a gaussian kernel, $w_k$ elements wide. The value of $c_{min}$ must be related to the contents of the kernel. In practice, it is a function of $w_k$ by giving it the value of the next-to-central kernel element.

Next, starting at a first maximum concavity position, all straight lines connecting this position with the other concavities are analysed. If the distance between the two concavities, measured along the shortest of the two contour segments, is shorter than a distinct length threshold $d_{min}$, the connecting line is discarded as a potential cutting line, as illustrated in figure 1. This is done to prevent separation of objects into fragments smaller than the smallest ones normally encountered with human chromosomes at the magnification employed.

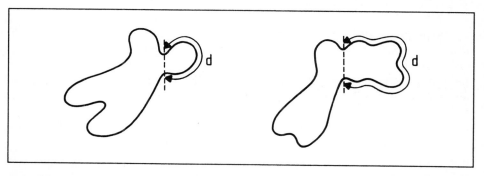

**Fig. 1.** The separating line in the left object is discarded, because the distance $d$, measured along the shortest of the two contour segments, is shorter than a distinct length threshold $d_{min}$, as opposed to the right object, where $d$ is sufficiently long; note that in both objects $g > g_{min}$ is assumed.

The value of $d_{min}$ should be related to the typical width $w_{typ}$ of all chromosomes in the metaphase. Note, however, that $w_{typ}$ must be derived from the metaphase image itself, but deriving it would require the image to be already properly segmented! So, by default, the value of $w_{typ}$ is derived from the distance histogram of the primarily segmented metaphase image. It is assumed equal to the distance at which the number of occurrences decreases most steeply. The intention of using $d_{min}$ is to prevent that parts of an object are cut off that are smaller than the smallest chromosome. A connecting line is also discarded if it contains more than a predefined number $n_{max}$ of pixels with a grey-value darker than the position of the object mode in the image grey-value histogram.

The optimal cutting line now is assumed between that pair of concavities that shows the highest value of $g$. This quantity $g$ is the ratio between the squared distance between the two concavities, measured along the shortest of the two contour segments, on the one hand, and the squared length of the straight connecting line on the other hand. In order to enhance cutting at very narrow sites, the denominator of this ratio is decreased by a constant $s$, taking care, of course, that the result remains positive. Again $g$ must exceed a predefined threshold $g_{min}$ with the intention to prevent cutting lines hardly shorter than the piece of contour they shortcut.

If a candidate pair of concavities has been found, their positions on the contour are optimized such that the distance between them is shortest. This process can be visualised as letting a rubber band connecting the original concavity positions slip along the contour. The optimization of the origin and target positions of the cutting line may, of course, interfere with the preceding selection of these points. However, applying it to every pair of concavities is considered to provide only marginal gain compared with its cost in computing time.

If no candidate pair of concavities shows a value of $g$ greater than $g_{min}$, the situation may be that of two or more chromosomes lying alongside each other. This situation is detected – though not specifically – by requiring that the minimum enclosing rectangle (MER) of the object has a width of at least twice $w_{typ}$. The process is repeated, but now the denominator of $g$ is multiplied by $\sin(2\phi)$, $\phi$ representing the angle between the longest axis of the object's MER and the connecting line. By doing so, the value of $g$ increases if the angle between the connecting line and one of the sides of the object's MER decreases.

Since single chromosomes rarely show very sharp bends, "strong" concavities in an object contour are forced to act as origins for cutting lines. To this end, first the local curvature maximum is analysed syntactically, by looking at the chain code strings representing the approximately straight contour segments immediately preceding and following. The difference between the average chain code in each of these strings is a measure for the complement $\alpha$ of the angle between the two "straight" contour segments. Note that the anisotropy of the square pixel grid makes it a far from perfect angular measure: straight line segments parallel to the grid axes are described with half the angular precision of their diagonal counterparts. A concavity is considered "strong" if $\alpha$ is sharper than a predefined threshold $\alpha_{min}$. Then a target position is assumed at minimal distance from the origin concavity on the "opposite" contour segment, i.e. skipping $d_{min}$ contour elements on both sides of the origin. The target position need not be a concavity, in this case.

Finally, the two contours that would result from cutting the object are analysed. The object will only be cut if the MER's width of one of them is larger than a distinct threshold value $w_{min}$. Applying this condition seems redundant in comparison with the previous requirement that origin and target should be $d_{min}$ contour elements apart. However, this condition is intended to prevent that well-separated chromatids are cut off a chromosome, at least if they are sufficiently straightened. Both size criteria will be of little help, of course, if biological fragmentation of chromosomes can be expected.

## 3. Results

The algorithm described in the preceding section has been applied to a set consisting of 20 metaphases selected by a cytotechnician as being representative of the best metaphases on 4 bone-marrow slides of one patient. Another 15 metaphases were taken more or less at random from the same 4 slides. The metaphase images were captured at the department of Human Genetics using the Magiscan-2, at the magnifications mentioned earlier, and transferred through a computer network to the VAX 11/750 computer system of the Medical Informatics Department. On the latter system, the algorithm was developed. An example of its results is given in figure 2.

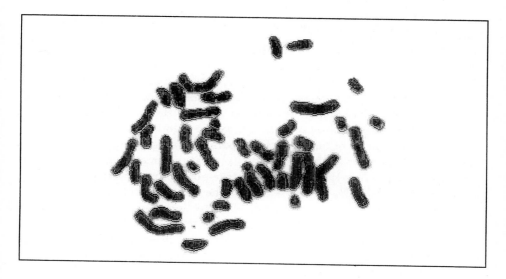

**Fig. 2.** Result of the segmentation algorithm on a typical bone-marrow metaphase.

In order to obtain some guidance in "tuning" the algorithm, an evaluation scheme has been worked out [7]. Essentially, the tuning consists of selecting which of the various features presented in the preceding section should be applied as criteria, their order, and their value. The selection process has been largely heuristic, i.e., it has by no means been performed exhaustively, because of the enormous number of possible combinations. Besides, the learning set consisted of the same 35 bone-marrow metaphases as the test set, plus 15 images of "typical problems" in bone-marrow metaphases. In the preceding section is described which features have been used as criteria, and in which order they were applied. After "optimization" the criterion values employed in obtaining the results presented here were the following:

- thresholding: adaptive;
- global offset for the threshold: $\delta T_a = 1$;
- width of the smoothing kernel: In practice, an odd value related to $w_{typ}$ is taken: $w_k = 2 * w_{typ} + 1$, with a maximum of 21, for practical reasons;
- minimum curvature $c_{min}$: the value of the next-to-central gaussian kernel element, i.e. $\binom{w_k}{w_{typ}}$;
- minimum distance between two concavities, measured along the contour: $d_{min} = 1.0 * w_{typ}$;
- maximum allowed number of pixels with grey-value darker than the object mode on a separating line: $n_{max} = 2$;
- constant for enhancing short separating lines: $s = 8$;
- minimum "shortcut ratio": $g_{min} = 2.5^2$;
- minimum angle for "strong" concavities:     $\alpha_{min} = \frac{\pi}{3}$;
- minimum width of the MER: $w_{min} = \frac{3}{5} * w_{typ}$;

In the evaluation scheme, the following quantities were used:

- $N_t$: the total number of objects, obtained visually;
- $N_o$: the number of overlapping objects, obtained visually;
- $N_f$: the number of "foreign" (non-chromosome) objects, obtained visually;
- $N_p$: the number of objects found after primary segmentation;
- $N_a$: the number of objects separated by the algorithm;
- $N_{FP}$: the number of false-positive separations among $N_a$;
- $N_{FN}$: the number of (false-negative) separations missing among $N_a$;
- $N_{TP}$ $(= N_a - N_{FP})$: the number of true-positive separations among $N_a$;
- $N_i$ $(= N_t - N_o - N_p = N_{TP} + N_{FN})$: the ideal value of $N_a$;
- $N_c$ $(= N_{FP} + N_{FN})$: the number of corrections required to arrive from $N_a$ at $N_i$;
- $q$ $(= N_c/N_i)$: the ratio between the number of interactions required *after* application of the algorithm, and the same number *before* application;
- $r$ $(= 1 - q)$: the "reduction ratio";

From the relations between these quantities follows:

$$q = \frac{N_c}{N_i} = \frac{N_{FP} + N_{FN}}{N_a - N_{FP} + N_{FN}}$$
$$\Rightarrow r = 1 - q = \frac{N_a - 2N_{FP}}{N_a - N_{FP} + N_{FN}}$$

The "reduction ratio" $r$ is preferred as a quality measure over the ratio $N_{TP}/N_i$, because it better reflects the reduction in the required number of interactions caused by application of the algorithm. It makes sense to put the value of $r$ in perspective. If, e.g., the number $N_p$ is well-optimized, the number

$N_a$ will be reduced, but $N_i$ also. Assuming $N_c$ constant, this will result in a much lower value for $r$, in spite of the fact that the overall result has remained the same.

As soon as the algorithm produces more false-positive cuts than true-positive ones, $r$ may even become negative. Then the necessary deletions outnumber the cuts correctly applied automatically, so it takes more effort to correct the results of the algorithm than it takes to apply all cuts manually.

Note that the number of objects $N_t$ in the metaphase image is not considered relevant in this scheme. "Foreign" objects like interphase nuclei and debris should not be excluded from the evaluation, because the chromosomes should be separated from them too. It would be even better if these objects were automatically eliminated, but that is not the primary aim of an algorithm for separating touching chromosomes. By including the "foreign" objects, a cut in one of them should also be considered false-positive. Overlapping chromosomes have not been taken into account in the evaluation of the algorithm, because their separation would require a different approach. For these reasons $N_i$ is taken as $N_t - N_o - N_p$.

**Table 1.** Results of the segmentation algorithm with the "optimal" parameter setting.

| Metaphases | $w_{typ}$ | $N_t$ | $N_o$ | $N_p$ | $N_a$ | $N_{FN}$ | $N_{FP}$ | $(N_i - N_c)/N_i$ | $= r$ |
|---|---|---|---|---|---|---|---|---|---|
| 15 BN..X: | | 694 | 12 | 337 | 284 | 78 | 17 | 250/345 | 72% |
| Average: | 14 | 46 | $= 1$ | $+22$ | $+19$ | $+5$ | $-1$ | 17/23 | |
| 20 BN...: | | 891 | 14 | 622 | 215 | 63 | 23 | 169/255 | 66% |
| Average: | 11 | 45 | $= 1$ | $+31$ | $+11$ | $+3$ | $-1$ | 8/13 | |
| 35 BN...: | | 1585 | 26 | 959 | 499 | 141 | 40 | 419/600 | 70% |
| Average: | 12 | 45 | $= 1$ | $+27$ | $+14$ | $+4$ | $-1$ | 12/17 | |

The results are presented in table 1, for the selected and the randomly taken metaphases, respectively. The differences between the two sets, consisting of 20 selected metaphases, and 15 ones taken at random, are immediately apparent: The primary segmentation produces about half the number of objects in the latter case, and about twice the number of cuts is required. In spite of this large difference in quality of the input metaphases, the tables show a fairly stable value for $r$. In general, the number of required interactions decreased by 419, from 600 to 181, after application of the cutting algorithm: a reduction with $r = 70\%$. However, there were 3 metaphases, for which the number of required interactions after application of the algorithm was higher than before.

**Table 2.** Results of the original Magiscan-2 segmentation algorithm, at maximum "zoom".

| Metaphases | "zoom" | $N_t'$ | $N_p'$ | $N_a'$ | $N_{FN}$ | $N_{FP}$ | $(N_i - N_c)/N_i$ | $= r$ |
|---|---|---|---|---|---|---|---|---|
| 15 BN..X: | | 683 | 297 | 7 | 380 | 1 | 5/386 | 1% |
| Average: | 1.49 | 46 = 20 | +1 | | +25 | –0 | 0/26 | |
| 20 BN...: | | 908 | 511 | 61 | 337 | 1 | 59/397 | 15% |
| Average: | 1.37 | 45 = 25 | +3 | | +17 | –0 | 3/20 | |
| 35 BN...: | | 1591 | 808 | 68 | 717 | 2 | 64/783 | 8% |
| Average: | 1.43 | 45 = 23 | +2 | | +20 | –0 | 2/22 | |

An attempt has been undertaken to compare these results with those obtained using the original segmentation software (table 2). As this software is intended for interactive use, it has been employed that way, by marking the individual chromosomes erroneously segmented in the primary step. Note that $N_t' \neq N_t$, because the metaphases have been framed differently in this experiment, including some isolated chromosomes and cell nuclei. (The primes indicate the results of the Magiscan-2 software as opposed to the algorithm presented here.) If it was left to function all by itself, the system often was not able to handle the situation, and a reboot was required. Even with interactive marking, results were poor: $N_p'$ turned out to be lower for all metaphases, and in 14 cases a negative value for $r$ resulted. The most likely cause of the poor performance is in the "light" global segmentation threshold employed, often resulting in large clusters of chromosomes taken as one object. Next, the local adaptation of this threshold fails, because of the large area involved. Besides, the original Magiscan-2 algorithm practically prevents false-positive separations, which — according to the theory of ROC-analysis — leads to a much higher false-negative rate, unless the segmentation is substantially better.

It should be noted, however, that a comparison like this is of limited value, because the original software was intended for different material, and for a different interaction pattern.

## 4. Discussion

For tuning the algorithm, it is absolutely necessary to have a representative set of as many as possible reference images. In practice, between 20 and 50 images is considered a minimum. On the Magiscan-2 in Leiden, only provided with a double floppy disk system and no hard disk, only one full image can be contained on a single diskette, and it takes more than two minutes to load the image into memory. This precludes the use of "batch mode" evaluation runs. Added to which, the facilities on the Magiscan-2 for program development are rather frugal, compared with those found on a general purpose computer such as the VAX. The availability of a large disk, a line-printer, and a Pascal compiler with cross-reference facilities are of great help in development, debugging and evaluation, if the system is also equipped with a frame-buffer and a display with comparable performance as the Magiscan-2. For this reason, most of the development presented here is done on the Medical Informatics department's VAX, rather than on the Magiscan-2 itself.

The problems of separation can be generalized into the numbers of wrong decisions taken over whether to cut and where. The greatest improvement can be obtained by carefully tuning the criteria for selecting the starting and target point of the cutting line. But even if the algorithm had been more rigorously optimized, it would have still retained certain limitations, since perfection is probably unobtainable with objects demonstrating the degree of variability such as chromosomes. So for example, the characteristic inward bend is not always present, resulting in undersplitting, and chromosomes containing many bends will tend to be fragmented, e.g., at their centromere position. If all centromeres were sufficiently darkly stained, this type of fragmentation could be prevented, by detecting a cut through a well stained region. More substantial improvements may be expected by incorporating other techniques into the algorithm.

The algorithm also has been implemented on the Magiscan-2 itself. In this implementation the interaction pattern is different from the one employed in the original segmentation software of the Magiscan-2. This algorithm does not use a-priori knowledge about the number of chromosomes in a cluster, so it only requires amendments to its results in the form of deleting false-positive and applying false-negative cuts. It still remains an open question whether the simpler interaction of indicating the number of chromosomes in each cluster would not result in a better overall performance. In order to answer this question a more profound analysis of the interaction pattern is required. In this analysis, the idle operator time will be an important factor.

Suppose the algorithm itself would not require any more interaction, i.e., $r = 1$ would be realised for all metaphases, which is highly improbable. Even then, a serious reduction in operator time can only be achieved if the remaining necessary operator intervention is concentrated in short periods, separated by long intervals, like in most automatic metaphase finders. Such an operating mode imposes severe demands on the metaphase relocation precision and on the reliability of the high-resolution autofocus, because relocation then must be performed with the ($\times 100$) oil immersion objective on the microscope. Moreover, a very robust implementation of the automatic segmentation is also required in this unattended operation mode.

The necessary corrections to the automatic segmentation then must be applied next, during the actual karyotyping step, to a series of temporarily stored digital images, e.g., the metaphases of a slide. Such an operating mode would not only require that a series of images consisting of more than 10 Mbyte can be stored on-line, but also that the display quality of these digitized images closely approximates that of their microscopic originals. An additional advantage of this operating mode would be the possibility of multiple-metaphase karyotyping.

# References

1   Graham J, Taylor CJ. Automated chromosome analysis using the Magiscan Image Analyzer. Analytical and Quantitative Cytology 2:237-242 (1980)
2   Graham J. Automation of routine clinical chromosome analysis I. Karyotyping by machine. Analytical and Quantitative Cytology 9:383-390 (1987)
3   Graham J, Pycock D. Automation of routine clinical chromosome analysis II. Metaphase finding. Analytical and Quantitative Cytology 9:391-397 (1987)
4   Vossepoel AM, Smeulders AWM, van den Broek K. DIODA: Delineation and feature extraction of miroscopical objects. Computer Programs in Biomedicine 10:231-244 (1979)
5   Rutovitz, D., Expanding picture components to natural density boundaries by propagation methods. The notions of fall-set and fall- distance. Proc. 4th Int. Joint Conf. Pattern Recognition (Kyoto, Nov.7–10, 1978) pp.657-664.
6   Smeulders AWM, Vossepoel AM, Vrolijk J, Ploem JS, Cornelisse CJ. Some shape parameters for cell recognition. In: Gelsema ES, Kanal LN (eds.)   Pattern Recognition in Practice (North-Holland, Amsterdam 1980) pp.131-142.
7   Vossepoel AM. Analysis of Image Segmentation for Automated Chromosome Identification. University of Leiden: Doctoral dissertation (1987)

# Model-Based Contour Analysis in a Chromosome Segmentation System

Q. Wu, J. Snellings, L. Amory, P. Suetens, A. Oosterlinck

## Summary

In this paper a recent investigation of using a model-based contour analysis technique to recognize and resolve composite chromosomes is described. Based on the fact that much of the information needed in delineation of metaphase chromosomes is outlined by the structural configurations of the chromosome contours, and if it is further assumed that most essential characteristics of such information are maintained by a polygonal representation resulting from a least square linear fit of the contour curves, the task of recognition and resolution of composite chromosomes can be largely tackled by a careful analysis of different object contour configurations using the model knowledge. The technique was tested on a set of G-banded metaphase images and the results are presented.

## 1. Introduction

Chromosome segmentation remains a difficult task in automation of chromosome analysis. After considerable early attempts [1-5] to solve the problem by using various image segmentation techniques, the expected degree of automation is far from reached and in existing systems much time-consuming human operation is still needed in image editing sessions, mainly for interactive delineation of composite (overlapping and touching) chromosomes that frequently occur in metaphase images.

Recently however, renewed interests arose in automatic resolution of composite chromosomes, by exploiting more task-specific knowledge in developing segmentation algorithms. From the work of Ji [6,7], in which very encouraging results were obtained, it is shown that using the knowledge based upon a careful study of photometrical and morphological models of composite chromosomes, and using intelligent search based control strategies in combining partially successful techniques to resolve various kinds of composites could result in much more capable systems than previous ones. On the other hand, progress has also been made in using available knowledge in an interactive segmentation system by a fall-set segmentation technique [8] to reduce the number of interactions required.

In this paper our recent investigation of using a model-based contour analysis technique to recognize and resolve composite chromosomes is described.

**Automation of Cytogenetics**   Editors: C. Lundsteen  J. Piper
© Springer-Verlag Berlin Heidelberg New York 1989

Based on the fact that much of the information needed in delineation of metaphase chromosomes is outlined by the structural configurations of the chromosome contours, and further assuming that most essential characteristics of such information are maintained by a polygonal representation resulting from a least square linear fit of the contour curves, recognition and resolution of composite chromosomes were attempted by a model-based analysis of object contour configurations. The study bears analogy to Ledley's early work [2] on decomposing homogeneously stained chromosomes by boundary analysis, and to the early work of the Leuven group [4] in using a linear piecewise polygonal appproximation for contour shape ananlysis, but differs in that the polygonal approximation of contour curves is made by a least square line fit procedure to minimize the coarse approximation error, and the primitive segments are defined and derived differently. Besides, a more extensive study of morphological models of different metaphase objects in terms of their contour configurations has been made.

In the following we begin by introducing some preprocessing steps used, and then describe in some detail the technique investigated in this study. Next the results from our experiments with metaphase images using this technique are presented. This is followed by a discussion about some of the benefits and potential use of the technique.

## 2. A Model-Based Contour Analysis Technique

### 2.1. Preprocessing

In this paragraph we briefly introduce some of the necessary preprocessing steps by which an intermediate result in terms of a binary segmentation mask can be obtained and used as a basis for proceeding contour analysis operations.

Presently for initial segmentation of metaphase images, a grey value histogram-based global thresholding operation is used. Following this a number of binary image processing techniques that are specially suited to enhance and improve the initial segmentation are further applied. These techniques were originally adopted from a software package developed in Delft University of Technology [9], and are well integrated into our environment. The following is a summary:

1. Removal of small non-chromosomal stain-debris by a couple of erosions followed by a propagation. The number of erosions must be enough to eliminate the small debris, but too few to remove the smallest chromosomes. The propagation then restores the original proportions and shapes of the remaining objects.

2. Removal of large non-chromosomal objects (interphase nuclei) by a number of erosions that are enough to eliminate all objects except those too big to

be a chromosome. After propagation these large objects are put away by taking an exclusive or-ing with the original binary input image.

This operation is made optional in the preprocessing however, because the situation may occur that the procedure tends to eliminate large composites of chromosomes segmented as one object, or in case of chromosome(s) touching a nucleus. In a later section (2.4) a method to erase the nuclei without the above mentioned problems is described, based on the contour analysis technique.

3. Holes within the objects are closed by a propagation from the edge of the image into the background mask (background is filled) after which the remaining parts belong to the objects.

4. Objects connected to the edge of the image are removed by a propagation from the edge into the object mask (border objects only) followed by an exclusive or-ing with the original binary input image.

5. Separation of slightly touching chromosomes by a few erosions after which the background skeleton is computed in order to create a dividing boundary between the split chromosomes.

The above techniques have been proven to be quite efficient and effective for our preprocessing purpose of the initially segmented images. To this end, a preprocessed binary segmentation mask is obtained and from this we can proceed to describe how the contour description and its polygonal approximation are derived, and the model-based contour analysis technique at work.

## 2.2. Contour Description and Polygonal Approximation

On a binary segmentation mask image, applying a contour following algorithm results in a contour curve description for each object segment in terms of its chained pixel coordinates stored in a cyclic list of record data structure. The contours are closed planar curves, and a parameterization is made, by following the curve clockwise and assigning to each successive pixel a coordinate pair:

$$x[n], y[n]; n = 1...N$$

where N is the number of pixels of the contour. This representation was chosen for a number of advantageous reasons in the computation of contour features on which our contour analysis is based. In order to carry out a structural analysis of the contours in the image, especially when a large number of objects are present, for efficiency reasons it is necessary to perform some kind of data reduction, preserving only the characteristic information. A number of curve fitting techniques exist for the approximation purpose using for instance polynomial or spline functions [10]. In our system a first order least square line fit was used, for simplicity and efficiency consideration. Thus contours are approximated by polygons, each one of the polygonal sides being the least square line fit of an arc segment, which is given in parametric form (x[n],y[n]). Least square fit in the sense that the sum of squared distances between the line and each of the points of the arc segment is minimised. It is shown [11] that such a minimisation can be obtained by:

$$x = ax \cdot t + bx$$

$$y = ay \cdot t + by$$

where

$$ax = \frac{\sum_{j=1}^{n} (j - <j>) \cdot x[j]}{(n^3 - n)/12}$$

$$bx = \frac{\sum_{j=1}^{n} x[j]}{n}$$

and

$$ay = \frac{\sum_{j=1}^{n} (j - <j>) \cdot y[j]}{(n^3 - n)/12}$$

$$by = \frac{\sum_{j=1}^{n} y[j]}{n}$$

with the term $<j>$ denoting the mean value of $j$, i.e.,

$$<j> = \sum_{j=1}^{n} \frac{j}{n}$$

This algorithm has the advantage that it can be implemented recursively without the need for any test whether the contour $y = f(x)$ or $x = f(y)$ is single valued in the interval considered. Another interesting aspect for treating highly irregular contours using this approach is that no breakpoints need to be given as in high order approximation.

Polygonal approximation of the contour curve using the least square line fit can be achieved through scan along algorithms or hop along algorithms [12]. A scan along algorithm treats the points one by one, increasing the length of the line segments as long as some error criterion is satisfied. A hop along algorithm treats subarcs of the curve, splitting them up if the error criterion fails. These can be easily implemented in a recursive way. In this implementation the hop along algorithm is used for the polygonal approximations based on an error norm criterion. In the error norm criterion the sum of the squared euclidian distances between the contour curve points and the line fit is taken for evaluation. The error norm is defined as:

$$E = \sqrt{\sum_{i=1}^{n} (x[i] - ax \cdot i - bx)^2} + \sqrt{\sum_{i=1}^{n} (y[i] - ay \cdot i - by)^2}$$

and the criterion is to keep E below a certain threshold value. At the moment a value of 1.2 image grid distance is applied. It should be noted that in general the procedure needs to be complemented with a merge step, in which those almost collinear neighbouring segments are joined together.

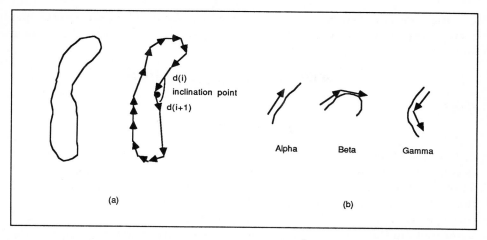

**Fig. 1.** (a) The fitting tangent vectors and the angle in between. (b) Three different elementary segments used to describe the contour shapes of the chromosomes.

## 2.3.  Contour Features and Pattern Primitives

From such a polygonal contour representation we can readily extract a number of important features, such as curvature and tangent directions. Consider a strict mathematical formulation of the tangent and curvature in the continuous case, the contours are represented by curves:

$$x = g(s)$$

$$y = h(s)$$

g(t) and h(t) being periodic continuous functions with the regular parametrization (s = arclength along the curve between (x(0),y(0)) and (x(s),y(s)). Then the angle between the tangent and the x-axis is given by:

$$\Phi = \arctan \frac{dh(t)/dt}{dg(t)/dt}$$

And the curvature is defined by:

$$\kappa = \lim_{\triangle s \to 0} \frac{\triangle \Phi}{\triangle s}$$

By analogy, in discrete case, ax and ay derived from the least square line fit can be considered as the components of the fitting tangent vector. The angle between two consecutive fitting tangent vectors is a fairly relevant measure for the curvature in the neighbourhood of a point called inclination point which separates the two segments (figure 1(a)).

Here we introduce the convention that this angle is negative when the curve is locally concave and positive in the opposite case. The global curvature between two fitting segments is therefore defined as:

$$\kappa = \angle(d_{j+1}, d_j) = \Phi_{j+1} - \Phi_j$$

where

$$\Phi_j = \arctan \frac{ay_j}{ax_j}$$

In this convention we may associate a negative curvature with the local event that a concave contour segment is encountered. To this end we can define and derive a set of structural pattern primitives from the polygonal description of the contours and the features extracted. On an intuitive basis one can consider chromosome segment contours as a family of shapes, substantially composed of elements from the structural pattern primitive set. As shown in figure 1(b) the chromosome contour shapes can be described by three different contour curve segments: let us call a straight segment an alpha, a convex segment a beta and a concave segment a gamma. These pattern primitives together with the features extracted from the polygonal contour description for each object segment are kept in a list of cyclic records other than the chained curve description used in the contour following, for later recognition and resegmentation purposes.

In general the morphology of the chromosome shapes suggests that possible protrusion or intrusion of the chromosome segments is normally indicated by the occurence of the gamma segments. This heuristic fact has lead us to follow the present approach in which contour analysis is carried out mainly based on gamma segments plus a geometrical analysis of the local spatial contextual information, instead of a strict string grammar one. A number of geometrical invariant relationships can been abstracted in this approach, by which the contour configurations are characterized and serve as models. These models can be implemented as parsing rules that perform relational tests on the pattern primitives and the local spatial contextual information of the polygonal contour description.

## 2.4. Models of the Contour Configurations

In this section, the models of a number of different configurations of object segments are studied. Only typical configurations and their parsing criteria are to be considered here for sake of space, but without loss of generality as the others can normally be comprised of these typical ones in one way or another. A more detailed description about this can be found in [13].

1. Overlapping chromosomes: See figure 2(a).
   The cases of overlapping chromosomes generally consist of overcrossed composites and T-crossed composites. The overcrossed composite case is characterized by contours containing 4 gammas forming 2 by 2 pairs of

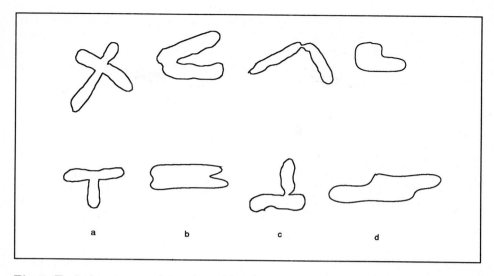

**Fig. 2.** Typical contour configurations. (a) Overlapping chromosomes. (b) Touching chromosomes with at least one strong gamma. (c) Touching chromosomes with a small touching zone. (d) Touching chromosomes with no strong gamma, but large touching zone.

"matching gammas" (the upper case, figure 3(a)). Two gammas are said to be matching if they satisfy the following conditions:

    a) Gamma1 and gamma2next have approximately the same direction, as well as the linkage between the two inclination points.

    b) Gamma1next and gamma2 have nearly opposite directions.

    c) The sum of the curvature of gamma1 and gamma2 is about $-180°$.

The T-crossed composite is shown in the lower case, figure 3(a). It is very difficult to determine the extent of occluded portions of the chromosomes from this kind of overlapping composites using the contour information only, as confirmed in the next section by the results of our experiments, in which the false negative and most recognized but unseparated overlapping composites are due to this configuration.

2. Touching chromosomes:

They frequently occur in metaphase images. Their subcases can be further categorized:

    a) Contours with at least one strong gamma: See figure 2(b).

    By strong gamma, we mean that its curvature is highly negative (a criterion between $-120°$ and $-180°$ is applied). It is often the case when chromosomes are touching at side. The search of the second gamma can be made by using a direction fitting method and the angle of the

loop. This is done by determining a direction of search by the strong
gamma, and finding the second gamma which minimizes the distance
between its position and the search direction (figure 3(b)). The angle
of the loop between two gammas is defined as the angle between the
segment following the first gamma and the second gamma.

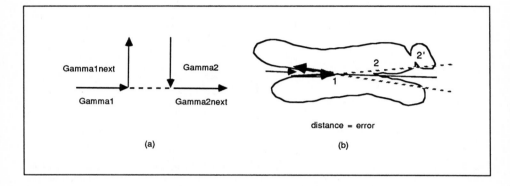

**Fig. 3.** (a) Matching gammas. (b) Direction fitting.

b) Contours with a small touching zone: figure 2(c)
   The definition of "small" is defined by comparing the distance between
   two gammas with the arm-width of the chromosome. A small touching
   zone generally means that the end of a chromosome is involved in
   the touching zone. When both chromosomes are touching at their
   ends, it is difficult to distinguish them from the centromere zone of a
   chromosome. Fortunately this rarely happens in practice.
c) No strong gamma, large touching zone: figure 2(d).
   These contours are more difficult to describe in terms of their gammas. The lower case is characterized by vectorial relations similar in
   nature to those characterizing the matching gammas, and value of
   the curvature (close to $-90°$). For the upper case, it can happen that
   the contour segment gives only one significant gamma (considerably
   negative). In such a case the object is parsed as being problematic for
   further resegmentation, with suggested hypotheses of either touching
   or bent chromosome(s). This is because from the polygonal description of the contours information about weak gammas tends to be lost
   due to the smoothing, making it difficult to differ such a configuration
   from the one of a bent chromosome.
3. Interphase nuclei and non-chromosomal stain debris:
   Interphase nuclei can be easily recognized as they contain no significant
   gamma segment, and their areas are large. On the contrary, stain debris

have very small area. Simply using this model knowledge in contour analysis, the problems mentioned in section 2.1 can be avoided in removing the nuclei from metaphase images.

For composites with more chromosomes involved, they are in fact the superposition of several typical configurations of which the individual structure of the gammas remains unchanged. This is the major advantage of working only with gammas for recognition, instead of with the complete set of the primitives. For the parsing criteria in the latter case would be enormously complex, even if the study is restricted to the composites of less than 4 chromosomes.

## 3. Results

The technique described in this paper has been implemented in C on a VAX workstation. Results from our experiments with a set of 14 metaphases are presented here. Some features of the test data are summarized in Table 1.

**Table 1.** Summarized features of the test data, number of metaphases tested = 14.

| Composite types | Overlapping chromosomes | Touching chromosomes |
|---|---|---|
| Number of composites | 13 | 78 |
| Successful recognition | 12 (92.3%) | 65 (83.3%) |
| Successful recognition and separation | 7 (53.8%) | 35 (44.9%) |
| Successful recognition with wrong separation | 0 (0%) | 2 (2.6%) |
| Successful recognition without separation | 5 (38.5%) | 28 (35.9%) |
| Wrong recognition (False positive) | 2 (14.3%) | 8 (11.0%) |
| Recognition failure (False negative) | 1 (7.7%) | 13 (16.7%) |

Example images at different processing stages are illustrated in figures 4, 5 and 6.

From the original metaphase images in figure 4, the object contour curves are extracted and drawn in figure 5. Using the model-based contour analysis technique described in the previous section, most of the composites of touching and/or overlapping chromosomes are recognized, and many are separated as well. Composites recognized but without separation are delineated by dashed

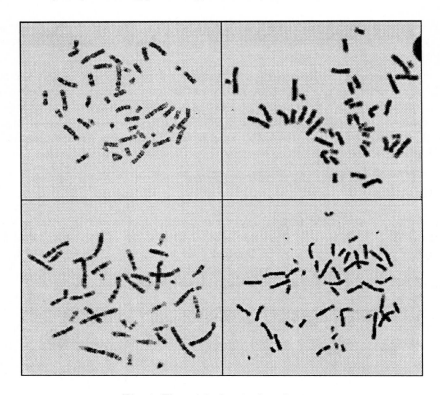

**Fig. 4.** The original metaphase images.

contours in figure 6. These are noted and put on a stack, subject to further resegmentation analysis.

Observing the test data, we notice that this use-contour-only technique results in quite a high successful recognition rate. The false negatives are associated with a T-crossing rather than overcrossing composite in the overlapping case, and mostly with the occurrences of small chromosomes forming elongated objects in touching cases. On the other hand the false positives in overlap recognition correspond to the special cases that clusters of small chromosomes exhibit contours similar to those of overcrossed chromosomes, and in touching recognition correspond to heavily bent chromosomes and chromosomes with high centromere contraction. The relatively high failure rate in finding plausible separation paths for those correctly recognized overlapping composites is mainly due to the difficulties in finding occluded parts of T-crossing composites, and in complex composites in which overlaps and touches closely join together the simple direction fitting search for a separation path turns out to be inadequate. In touching cases the smoothing effect of the polygonal approximation near touching zones tends to eliminate weak gammas which may contain important information about separation paths.

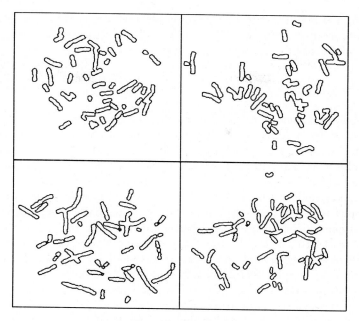

**Fig. 5.** The original contour curves.

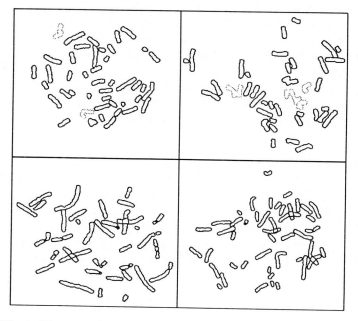

**Fig. 6.** The resultant contours after the model-based contour analysis.

During the experiments we also observed that the pattern recognition and corresponding separation decisions can be affected in a way by using different threshold offset values in initial segmentation.

## 4. Discussion

The main advantages of using this model-based analysis of object contours approximated by a least square polygonal representation are its simplicity, comprehensibility and effectiveness. In this approach an efficient analysis of different object configurations can be achieved, provided that important characteristics needed for the delineation of chromosome contours are maintained in such a polygonal approximation. The purpose of this study is to investigate the appropriateness of the technique as a computationally inexpensive means of hypothesis generation in knowledge-based chromosome analysis systems under development [14,15], in which object hypotheses are needed in choosing search based control strategies that combine various partially successful chromosome segmentation techniques to resolve different kinds of composites.

To substantialize this, consider following a proposed model-based contour analysis. Hypothesized composites which are unlikely to be resolved by use-contour-only methods could be verified and resolved by some particularly effective local image operations, e.g. a rethresholding or a larger number of erosions and propagations accompanied by background skeletonization for touching composites, and by using convex hull, valley following and skeleton analysis for overlapping composites. But these more elaborate and computationally more expensive procedures would only be called in on requests of a lower level hypothesis generation component. The small trial carried out in this study indicates that the model-based contour analysis technique does seem to provide reasonably good chromosome composite hypotheses for this purpose. The practice and realization of above conceived scenario however, will of course require much work ahead.

*Acknowledgements.* The authors would like to thank Claes Lundsteen and Jim Piper for their comments on the manuscript, and Luc Van Eycken for his advice on TEX coding. This work was supported in part by the Belgium national incentive program for fundamental research in Artificial Intelligence, initiated by the Belgium State - Prime Minister's Office - Science Policy programming, and by the Commission of the European Communities within the frame of its Medical and Health Research Program by project No. II.1.1/13.

# References

1.  Castleman KR, Melnyk JH. Automatic system for chromosome analysis–final report. JPL Document No 5040-30 Jet Propulsion Laboratory, Pasadena, California (1976)
2.  Ledley RS, Ruddle FH, Wilson JB, Belson M, Albarran J. The case of touching and overlapping chromosomes. In: Cheng GC, Ledley RS, Pollock DK, Rosenfeld A(eds) Pictorial Pattern Recognition (Thompson, Washington DC, 1968) pp.87-97
3.  Rutovitz D. Chromosome classification and segmentation as exercises in knowing what to expect. In: Elcock EW, Michie D (eds.) Machine Intelligence 8 (Ellis Horwood, London, 1977) pp.455-472
4.  Vanderheydt L, Dom F, Oosterlinck A, Van den Berghe H. Two-dimensional shape decomposition using fuzzy subset theory to automated chromosome analysis, Pattern Recognition 13:147-157 (1981)
5.  Vossepoel A. Analysis of image segmentation for automated chromosome identification, Ph.D. Thesis, University of Leiden (1987)
6.  Ji L. Intelligent splitting in the chromosome domain, Pattern Recognition, (in press, 1989a)
7.  Ji L. Decomposition of overlapping chromosomes, this volume (1989b)
8.  Graham J. Resolution of composites in interactive karyotyping, this volume (1989)
9.  ten Kate TK, Design and implementation of an interactive karyotyping program in C on a VICOM image processor, Engineer Thesis, Delft University of Technology, (1985)
10. Eklundh JO, Howako J. Robust shape description based on curve fitting, Proc. Int. J. Conf. on Pattern Recognition (1984)
11. Snellings J. Segmentation and feature extraction–A syntactical pattern recognition approach to the preprocessing of metaphase images, Engineer Thesis, Catholic University of Leuven, (1987)
12. Ramer U. An interactive procedure for the polygonal approximation of planar curves. Computer Graphics and Image Processing 4:244-256 (1972)
13. Amory L. Conversion and improvement of a syntactical pattern recognition package for the resegmentation of metaphase images, Engineer Thesis, Catholic University of Leuven, (1988)
14. Piper J, Towers S, Baldock R, Rutovitz D. Knowledge-based control of a chromosome analysis system. MRC Human Genetics Unit, internal report (1987)
15. Wu Q, Suetens P, Oosterlinck A. Toward an expert system for chromosome analysis, Knowledge-Based Systems 1:43-52 (1987)

Part VI

# Chromosome Classification

# On the Use of Automatically Inferred Markov Networks for Chromosome Analysis

E. Granum, M.G. Thomason, J. Gregor

## Summary

A structural pattern recognition approach to analysis and classification of chromosome band patterns is presented.

First discussed are aspects of band pattern representation. An operational method of processing profiles and deriving string representations, which focuses on the sequences of local band pattern features, is outlined. It is found that extra emphasis on transitions between bands gives the best results.

A method of syntactic analysis of band sequences by dynamic programming is described from a functional point of view. It concerns automatic inference of band pattern models (Markov networks) per class, and the use of these models for classification. The models may be so inferred that centromere position is encoded and used for centromere finding based on band pattern only. Combining classification and such centromere finding provides improved classification and fairly accurate centromere positions for unknown chromosomes.

Experimental results verify the potential of the method and suggest that karyotyping using this approach alone may give results comparable to other approaches, which use global features. It is proposed, however, that the best results may be achieved by combining approaches of global and local features.

## 1. Introduction

By their very nature, patterns to be analysed may call for statistical or structural descriptions or possibly both. Statistical descriptions are typically global measures such as size, while structural descriptions typically describe the relationships among the subpatterns that form the overall pattern (Thomason, 1987).

Chromosomes and their band patterns are examples of composite patterns, i.e., patterns with components related to both. A chromosome may be described by its size (e.g., length, area, integrated optical density (IOD), etc.) and centromere index, which are measures appropriate for statistical classification methods. The band pattern, however, includes local relationships of a structural nature.

Most attempts at automated chromosome classification have used global statistical properties only for band pattern description (Casperson et al., 1971, Piper et al., 1980, Granum, 1982, Piper and Granum, 1988), while only few rely on the locally dependent details, e.g., (Granlund, 1973, Lundsteen and Granum, 1979b,

**Automation of Cytogenetics**    Editors: C. Lundsteen J. Piper
© Springer-Verlag Berlin Heidelberg New York 1989

Thomason and Granum, 1987) or a mixture (Lundsteen *et al.*, 1981, Ten Kate, 1985).

This paper addresses structural aspects of the human chromosome G-band patterns and presents an approach dealing with such aspects (Thomason and Granum, 1986, 1987). The technique uses dynamic programming and includes facilities for automatically inferring a model of each chromosome type from learning data, as well as for classifying chromosome types using these models.

Previous smaller experiments on the Denver groups D and G (Thomason and Granum, 1986, 1987) have been expanded to all Denver groups and to the use of length information instead of a priori group assignments (Granum and Thomason, 1988). An overview of the results is provided. The technique assumed initially the orientation of the band pattern (p-q vs. q-p) to be known, and need not make any reference to the position of the centromere. However, the method includes a potential for coping with chromosomes of unknown orientation (Granum and Thomason, 1988) and for detecting centromere position on the basis of band pattern information only (Gregor *et al.*, 1988). These and other extensions are also presented with reference to some experimental results.

## 2.   Band Pattern Representation

### 2.1   Principle of Profile Processing

Various methods have been proposed for describing the *local* properties of the band pattern as derived directly from image data or from a density profile. Some approaches treat dark bands as independent entities, e.g., the profile model of Distribution Functions of Granlund (1973), and the bands detected in the image using the Laplacian filtering method of Visser (1981). Other approaches try explicitly to include the *sequential* nature of the band pattern, e.g., the Idealized Profile (Granum and Lundsteen, 1977, Granum, 1980) which may be derived from either the image or the profile, and the Band-Transition Sequences, BTS (Lundsteen and Granum, 1979a, 1979b). Most reported scientific experiments as well as most practical systems, however, rely on the simple profile.

The approach reported here uses the discrete version of the *idealized profile* obtained from a raw profile in a data base as illustrated in Figure 1. First, a simple 3 point smoothing (weights, 1,2,1) is performed to obtain a less noisy profile as shown in Figure 1b. A differential analysis is then carried out to establish positions of inflexions of the profile. The simple hypothesis is that positions of inflexions can be used as estimates of transitions between bands of different density. In most cases one local extremum will exist in each of the intervals between inflexions along the profile. As an estimate of the density of a band extending the entire interval between two successive inflexions, the density value at the local extremum in that interval is used to represent the "corresponding" band, as shown in Figure 1c. A *chromosome band* then comprises a sequence of one or more sample values; each such sample value is said to represent an *incremental band*. The result of this processing is an example of an idealized profile.

The differential analysis is implemented as a non-linear, iterative filter (Granum, 1980) based on a method of Kramer and Bruckner (1975). Special cases of more than one inflexion between successive extrema are made to enforce extra extrema. In practice, that corresponds to turning a "shoulder" of the profile into 2 bands, on the assumption that a shoulder is the rudiment of distinct bands. Figure 1 illustrates such situations, where the outermost bands at each end of the profile in Figure 1c are created in response to "shoulders" of Figure 1b.

The idealized profile has been shown to represent the essential band pattern features at least as well as the original profiles for visual classification (Lundsteen and Granum, 1978). Additional studies on "isolated" (i.e., out of cell context) visual classification of profiles showed an error rate of about 5% (Lundsteen and Granum, 1977), while visual classification of so-called BT-profiles that are based on a *subset* of the information of the idealized profile (Lundsteen and Granum, 1979a) showed an error rate of about 6% (Lundsteen and Granum, 1979b).

The idealized profile approximates the original profile with fewer density levels. However, there is no general agreement as to how many different density levels are required or relevant for description of chromosome band patterns. Density patterns vary between staining techniques and also within preparations. In the International System for Human Cytogenetic Nomenclature of 1985 (ISCN, 1985) only two levels are used to indicate the basic band pattern structure, while up to five levels are indicated in some schematic illustrations.

The principles of the profile processing above are in agreement with a statement in the ISCN (1985), page 13:

> "Each chromosome in the human somatic cell complement is considered to consist of a continuous series of bands with no unbanded areas. As defined earlier, a *band* is a part of a chromosome clearly distinguishable from adjacent parts by virtue of its lighter or darker staining intensity."

The interpretation of "clearly distinguishable" for digital profile data is possibly taken to the extreme in our case with the forcing of shoulders into two bands. However, while the ISCN (1985) illustrates examples of 400, 550 and 850 bands for a haploid set (22 autosomes + X and Y) our data base is characterized by 351 bands (Lundsteen *et al.*, 1980) and comprises thus on the average relatively contracted chromosomes. The corresponding band pattern enhancement is assumed justified, and we take this approach so far as to specify that all bands detectable in the idealized profile should also be preserved in any derived band pattern representation. The primary emphasis is on conserving actual transitions between bands rather than absolute or even relative density values.

## 2.2   String Representation of Band Patterns

An algorithm has been developed to map the idealized profile non-linearly into a profile of any specified integer number of levels. A linear scaling of the profile is computed and then systematically manipulated until all transitions have been restored. Figure 1d shows the result of mapping into 6 levels. This profile can be expressed directly as a sequence of symbols, *a string*, using the "alphabet" {1,2,3,4,5,6} as shown in Figure 1e. Each symbol represents an incremental band.

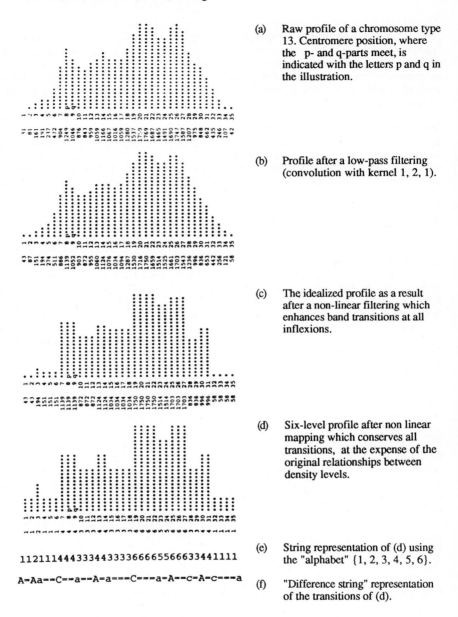

(a)    Raw profile of a chromosome type 13. Centromere position, where the p- and q-parts meet, is indicated with the letters p and q in the illustration.

(b)    Profile after a low-pass filtering (convolution with kernel 1, 2, 1).

(c)    The idealized profile as a result after a non-linear filtering which enhances band transitions at all inflexions.

(d)    Six-level profile after non linear mapping which conserves all transitions, at the expense of the original relationships between density levels.

11211144433344433336666655666633441111

(e)    String representation of (d) using the "alphabet" {1, 2, 3, 4, 5, 6}.

A=Aa==C==a==A=a===C===a=A==c=A=c===a

(f)    "Difference string" representation of the transitions of (d).

**Fig. 1.** Stages of profile processing

Band *transitions* are emphasized further in strings based on the sequence of the signed differences between successive incremental bands. Figure 1f shows a "difference string" of the same profile using the symbols (...,c,b,a,=,A,B,C,...) for differences of (...,-3,-2,-1,0,1,2,3,...). Using this representation, however, there is no guarantee that a string generated by an inferred network can be directly illustrated as a valid profile. Accumulated differences scanned along a generated string may not sum up to zero.

Pilot experiments have indicated that 4 to 6 levels are useful compromises and that difference strings are somewhat better for classification purposes than are strings of density levels. The actual representation used in terms of the number of levels, and the use of either density levels or differences is indicated in connection with each experiment.

# 3.  Material used in Experiments

## 3.1  Data Source

A database of 6985 digitized trypsin G-banded human metaphase chromosomes from 179 cells obtained from 12 normal individuals (Lundsteen *et al.*, 1980) is used in all the experiments referenced. On average there are about 300 samples of each autosome type, as overlapped and severely bent chromosomes are excluded. The chromosome images are digitised from photographs with a spatial resolution corresponding to 1/8 micron at specimen level. Density profiles are obtained by averaging perpendicular to the major axis of the chromosome area after segmentation. The same database has also been used in a number of other studies (Lundsteen *et al.*, 1981, Granum, 1982, Ten Kate *et al.*, 1983, Zimmerman *et al.*, 1986, Piper and Granum, 1988).

## 3.2  Datasets for the Various Experiments

Balanced datasets of difference string representations of just the autosomes (types 1 to 22) are used. The strings of each type are sorted by decreasing length. The middle 200 of each sorted list are extracted for use in the experiments, i.e., 4400 out of the 6985 chromosomes are used. Figure 2 lists examples of one string per chromosome type.

Thus for $j$=1 to 22, each dataset $d_j$ has 200 sample strings representing chromosomes scanned from the p-end towards the q-end; the centromere is not indicated. The ranges of lengths are given in Table 1. To create learning and test data, every other string in $d_j$ is copied into learning dataset $ld_j$; the remaining strings are copied into test dataset $td_j$. Hence, $ld_j$ and $td_j$ each contain 100 samples.

To allow studies of whether the string recognition is independent of p-q vs. q-p orientation, dataset $r_j$ is obtained from $d_j$ by reversing the strings and swapping upper case and lower case letters. Thus, the string AA=B=d would become D=b=aa. Strings in $r_j$ represent q-end toward p-end scans. Strings in $lr_j$ (or $tr_j$) are reversals of those in $ld_j$ (or $td_j$).

| Type | Sample String |
|------|---------------|
| 1 | A=A==a=A=a====A==a=====B====a==A==b==========E====e=====B==a==C==d====A==a====A==a==a |
| 2 | A==B=====a==A=a==A==a====A==a==A=a=A==a==A==a===C==a=A==b===C===b==A=b=A==c====a |
| 3 | A=B==a==B===b=====B==a==C==a=A==c==A====b=====B=======c=====A==a=a |
| 4 | A=A=a=====D==a==A=a==B===b====B==a===A=====c====B===c==A=b==a |
| 5 | A======E==c===B=b=A=a=A=a====C=====a==A==c==A=a=A===c======a |
| 6 | A=B==a==B===b=====C==a==B=c====B===b===A==b=A=b=====A=a==a |
| 7 | A====D==b==A==a====B==d=====E==d====D==d===A==b=====a |
| 8 | A===A===a=A==a==A===a=====C===a===C==e=========a |
| 9 | A=C==a=B===b===B==c=====D===c====B=b=A=c===Aa=a |
| 10 | A=A==a==A==a===C==b===D==d=====B==b==A=b===A=a=a |
| 11 | A=======C=a=A=a==A==c=======E=====e====A===a=a |
| 12 | A=====B==a==A==b=====E=====e========A==a=a |
| 13 | A=A=a===B==a===A==a===D===b=A==c==A==b==a |
| 14 | AA=a===B=a===D===c==A=a==A==c===A=a=a |
| 15 | B=A=a===D==a===A==d=====A==a=A==b=a |
| 16 | A=A===aA==a===E==e==A==a==A==a==a |
| 17 | A==Aa=====D==c=======D=====e==Aa=a |
| 18 | A=B==a==A=a==D===c====C==e==a |
| 19 | A=A=a=====E==e====A=a=A==a=a |
| 20 | A==C===a==C==d====A===b==a |
| 21 | AA=a====E===e======Aa=a |
| 22 | AA=a====E==e===A==a==A==a=a |

**Fig. 2.** Examples of "difference strings" of chromosomes. Each autosome type is represented by the sample in the learning set having highest alignment probability with the inferred networks.

**Table 1:** String length ranges and averages for the 200 samples of each chromosome type selected for the reported experiments.

| Type | Min($dj$) | Max($dj$) | Average($dj$) | Type | Min($dj$) | Max($dj$) | Average($dj$) |
|------|-----------|-----------|---------------|------|-----------|-----------|---------------|
| 1 | 69 | 105 | 85.6 | 13 | 37 | 45 | 41.6 |
| 2 | 69 | 97 | 82.2 | 14 | 36 | 44 | 40.6 |
| 3 | 58 | 81 | 70.0 | 15 | 35 | 42 | 39.4 |
| 4 | 56 | 73 | 65.0 | 16 | 32 | 37 | 35.5 |
| 5 | 55 | 70 | 63.1 | 17 | 31 | 37 | 35.0 |
|   |    |    |      | 18 | 29 | 35 | 32.8 |
| 6 | 52 | 71 | 61.3 |   |    |    |      |
| 7 | 49 | 63 | 56.2 | 19 | 25 | 31 | 28.8 |
| 8 | 45 | 57 | 51.5 | 20 | 25 | 31 | 28.5 |
| 9 | 43 | 55 | 49.4 |   |    |    |      |
| 10 | 43 | 54 | 49.0 | 21 | 20 | 25 | 23.4 |
| 11 | 43 | 54 | 49.2 | 22 | 22 | 27 | 25.6 |
| 12 | 43 | 54 | 48.9 |   |    |    |      |

# 4. Methods

## 4.1 Principles of the Dynamic Programming Inference Method

The mathematical definition and description of the inference method is presented in detail in (Thomason and Granum, 1986) and briefly in (Thomason *et al.*, 1986) and (Thomason and Granum, 1987). The method is outlined here from a functional point of view.

Consider difference strings as in Figure 2. Strings of the same type may be aligned and incorporated into one structural model by merging them where they are alike and allowing alternatives where needs be. Figure 3 shows an example of how this can be done in a systematic fashion by aligning and incorporating strings one by one. "Aligning" here means the computation of how to match, to substitute, to insert, or to delete symbols along a path in the given model to make it best accommodate the new string. "Incorporating" means modifying the model accordingly.

This model is a network with unique starting and termination nodes. By keeping track of how often a given arc has been included in the alignment of a new string, we obtain the frequency counts used to compute which of the alternative alignments makes the optimal changes to the network. Modifications to the nodes of the network with high frequency counts are more costly than modifications to low frequency nodes.

The alignments are computed by dynamic programming using a cost function based on these frequencies. The minimum costs of matching, substituting, deleting and inserting symbols as network modifications are computed for each new sample. At least one optimal alignment of minimum cost is guaranteed to be found; i.e., one for which the string will have the maximum achievable probability as a realization of the network. Associated with each optimal alignment there may be more alternative minimum cost alignments or "traces" through the network.

When the method is used for *inference*, an optimal trace is selected and incorporated for each learning sample in turn. The consequence is the reinforcement of known structure (update of frequency counts) or the learning of new structural details. The algorithm operates on a symbol by symbol basis; any modification may give rise to a considerable number of new variants of the strings represented by the network, i.e., strings that the network might output if used as a generator. This phenomenon is an essential property of an inference process, and it is illustrated in the example of Figure 3.

The inference technique thus automatically builds a constrained first-order Markov chain (a Markov network) from a set of strings through a data driven process. That is, no structure needs to be known a priori and noisy samples can be coped with. Common structure in the samples is found by the mechanism of aligning substrings that recur in the samples in about the same locations with the corresponding subpaths in the network. These subpaths gain higher and higher frequency counts. Hence, frequently aligned substrings obtain strong statistical support, and rarely seen subpatterns remain weakly supported.

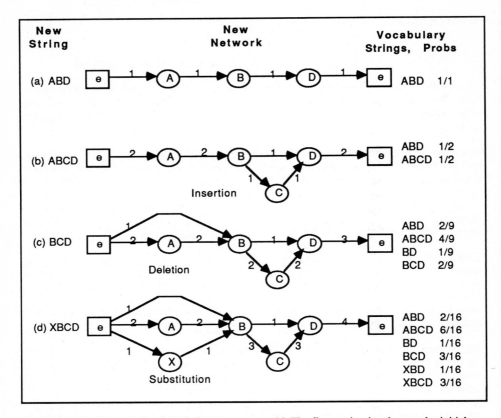

**Fig. 3.** Simple example of the inference process. (a) The first string is taken as the initial network. (b) The second string is accommodated by *insertion* of a new node. Notice the updated frequency counts on the arcs. (c) A *deletion* is needed to incorporate the third string. As a generator altogether four strings can be output. String BD is inferred. (d) Accommodation of the fourth string relies on a *substitution* operation. Yet another string XBD has now been inferred also as a possible, but not very likely output. Notice the most likely string ABCD includes the symbols most frequently observed.

When the method is used for *classification* purposes by aligning a string with a set of networks, the maximum alignment probabilities are computed. These values are in fact the probabilities with which the updated networks would generate that very string. An alignment probability is high if there is a good "match" between a network and the string. Conversely, if the string and a network "match" poorly, the alignment probability is low.

The network inferred from a given set of data depends on the ordering of the samples. We have found that ordering the samples by decreasing length gives as good results as other proposed orderings (Thomason *et al.*, 1986). An inferred network has an inherent asymmetricity related to the cost of an insertion operation. As network editing is computed on a symbol by symbol basis, the cost of inserting must be computed *either* from the statistics of the symbol succeeding *or* preceding the new symbol. The implementation uses the latter relationship ("inserting after") and the practical importance of the choice is possibly limited. This asymmetry is considered in some of the procedures suggested below (section 5.2).

## 4.2   Classification Rules

Three aspects of the classification procedure are considered for the initial experiments:

- a)   The selection of the candidate chromosome classes (types) to which a given sample must be compared.
- b)   The computation of the numerical value which relates a given sample to a class.
- c)   The criterion for final classification among candidate classes.

*a) Selection of Candidate Classes:*   Measures of size and centromere index can resolve at least 7 groups from the 22 types of autosomes. Some of our experiments assume the Denver groups known a priori and we test the ability of the structural models of bands to resolve by type within groups.

An obvious alternative to considering only one Denver group per sample is to try all types - here 22 classes. The computing involved would be extensive, as a full matching process is required for each sample-class combination. Since the length of a sample string is a reasonable reference for eliminating some networks from consideration, additional experiments are carried out as follows using an initial *"length-test"*. For each sample a set of candidate classes (networks) is selected by comparing the length of the sample to the range for each type (Table 1). Each class which according to this range could include the sample is a candidate class. Note that we thereby have a complete classification procedure using absolute length and band pattern information only.

*b) Sample-to-Class Relation:*   The distribution of alignment probabilities of learning data sets matched to their own networks vary considerably from type to type. The consequence is that a relatively good match with network 6 may yield a lower probability than a relatively poor match with network 10. Hence, a normalization within the alignment probabilities of each network may be required.

A proper normalization should be related to the distribution of probabilities of all the samples, which could be generated by the network. This distribution is very hard to get in full or approximate terms; therefore, the alignment probability of the most likely string of a network is used as a maximum reference probability (MRP) of that network. In some experiments we investigate and compare:

- i)   The direct use of the *absolute* alignment probability as the value of a string-to-network match.
- ii)   The use of a *relative* alignment probability, which is computed as the ratio between the absolute alignment probability and the MRP for the same network.

*c) Final Classification:* Classification according to the network yielding the highest (absolute or relative, respectively) alignment probability is used in all experiments. No ties have occurred.

## 4.3  PQ-Orientation of Strings

Potentially the performance of the method may depend on the orientation of the strings (p-q vs. q-p). The asymmetric property of the network might give rise to some orientation dependent results. In order to investigate the impact of variations due to orientation dependencies, some experiments are repeated for strings in reverse orientation.

## 4.4  Inference of Networks and Network Characteristics

For the individual experiments the appropriate number of networks are inferred, each in one run of a learning data set per class. No further processing is applied to the networks. Network $ndj$ is inferred from learning data set $ldj$, and network $nrj$ is inferred from learning data set $lrj$.

Typical network characteristics are exemplified in Table 2. The number of states excludes certain ("empty") states created for convenience in book-keeping during string alignments. The number of paths in a network from the starting node to the termination node is the number of different realizations of node sequences of the underlying Markov chain as established by the inference process. Since this number is very large for networks inferred from long and noisy strings, we give its logarithm in Table 2. The path entropy for *all* the realizations of a network used as a stochastic generator of node sequences is one fundamental characteristic of the probability distribution of these realizations. Let $p_i$ be the probability of realization $s_i$ of network $n$. Then the entropy is

$$H(n) = -\sum_i p_i \log(p_i)$$

The MRPs of the networks are the probabilities of generating the most likely string of each network. These values are used for normalization of alignment probabilities to relative values in some experiments.

Network characteristics are important means of monitoring the method during development, and many more are used. For more details see (Thomason and Granum, 1986, 1987, Thomason *et al.*, 1986).

# 5. Using the Markov Models for Classification etc.

In the following discussions we are looking into what the inferred Markov models may offer, by gradually introducing them in more and more comprehensive contexts. Some of their possibilities have been verified experimentally while others are suggestions which must undergo further investigations to prove their value.

The major part of the discussion is assuming isolated classification. That is, cell context dependent information like length normalization and expected number of chromosomes per class is not taken into account.

**Table 2.** Characteristics for the networks used in the experiments reported in sections 5.1 and 5.2. See text for comments.

| Network | No. of States† | log(no. paths) | Path Entropy | MRP |
|---------|----------------|----------------|--------------|-----|
| nd1/nr1 | 536/543 | 187.6/189.8 | 67.3/67.3 | 1.99e-13/--- |
| nd2/nr2 | 407/438 | 186.3/160.5 | 62.7/63.5 | 3.08e-13/--- |
| nd3/nr3 | 477/460 | 160.7/153.0 | 57.0/57.8 | 1.97e-10/--- |
| nd4/nr4 | 447/403 | 144.6/132.8 | 53.5/52.9 | 2.62e-10/--- |
| nd5/nr5 | 368/389 | 123.2/126.1 | 49.2/49.6 | 1.76e-09/--- |
| nd6/nr6 | 420/406 | 135.2/140.0 | 48.4/47.8 | 1.77e-09/--- |
| nd7/nr7 | 327/307 | 107.3/98.5 | 38.9/39.3 | 6.41e-08/--- |
| nd8/nr8 | 342/310 | 111.0/97.6 | 41.7/41.6 | 2.01e-08/--- |
| nd9/nr9 | 312/313 | 98.2/99.8 | 39.6/39.8 | 1.01e-08/--- |
| nd10/nr10 | 251/277 | 84.1/94.5 | 35.6/37.7 | 3.78e-08/--- |
| nd11/nr11 | 270/280 | 88.1/90.2 | 32.7/33.5 | 7.22e-07/--- |
| nd12/nr12 | 295/343 | 95.2/110.9 | 36.3/37.3 | 6.48e-08/--- |
| nd13/nr13 | 242/230 | 82.1/75.7 | 29.7/29.2 | 5.24e-06/--- |
| nd14/nr14 | 235/234 | 74.8/76.7 | 32.8/32.8 | 2.17e-07/--- |
| nd15/nr15 | 341/253 | 104.1/77.9 | 33.0/32.0 | 1.53e-05/--- |
| nd16/nr16 | 234/234 | 76.5/79.7 | 27.0/27.1 | 1.57e-05/--- |
| nd17/nr17 | 236/247 | 68.5/74.0 | 29.9/29.6 | 1.04e-06/--- |
| nd18/nr18 | 216/228 | 70.7/69.4 | 26.7/26.5 | 1.69e-05/--- |
| nd19/nr19 | 130/127 | 47.9/49.8 | 19.2/20.0 | 1.03e-04/--- |
| nd20/nr20 | 203/174 | 63.8/54.8 | 23.5/23.2 | 2.12e-04/--- |
| nd21/nr21 | 117/109 | 39.4/37.9 | 15.8/15.7 | 1.17e-03/--- |
| nd22/nr22 | 146/164 | 47.2/54.8 | 21.1/22.4 | 9.21e-05/--- |

† States used for bookkeeping purposes in a network are not counted.

## 5.1   PQ-Orientation Known

The simplest context for our experiments is provided when pq-orientation of the band patterns is known and no reference is made to the centromere. The following experiments (Granum and Thomason, 1988) use autosomes only and the band patterns are represented by 6 level difference strings.

*Within Denver Group Classification.* As classification into Denver groups may be carried out with only a few pct. errors using global features, it is of interest to see how many additional errors are made within Denver groups using the local band pattern structures as captured by the inferred networks.

Experiments show 8% error rate for absolute alignment probabilities and 6.4% for relative alignment probabilities. The latter results were also obtained with considerable lower variation in performance over the types.

*The 22-Class Experiments.* Testing each string (chromosome) against all 22 networks is very time consuming and not very relevant. Just the use of the unnormalised length measurement can reduce the relevant number of classes (networks) to consider per input string to about 4 on average.

Using this "length-test" instead of Denver groups reduces the recognition rate by 1-2%. Error rates are thus 9.8% for absolute alignment probabilities and 7.7% for relative alignment probabilities (Granum and Thomason, 1988).

*Isolated classification; all types*
*6 level difference strings; average error rate per type ($\pm$ std.dev.)*

| PQ-orientation | Pre-classification | Absolute alignment | Relative alignment |
|---|---|---|---|
| Known | Denver groups | 8%  ($\pm$7.1) | 6.4% ($\pm$4.3) |
| Known | Length test | 9.8% ($\pm$4.6) | 7.7% ($\pm$4.4) |
| UNknown | Length test | 13%  ($\pm$ 5.4) | - |

## 5.2   PQ-Orientation Unknown

The pq-orientation may typically be obtained by finding the centromere using image analysis procedures and assigning the terminal nearest to the centromere as the p-end. Assuming that centromere errors are positioned randomly along the chromosome, the orientation will only be wrong half as often as the centromere. That is, by using results of Piper and Granum (1988) about 4% pq-orientation errors may be expected on these data.

An alternative way of getting the orientation by using the Markov networks is to align each string also in its reverse orientation or to have individual networks for both orientations of each chromosome type, i.e., in our case we would have 44 classes: 1-pq, 1-qp, 2-pq, 2-qp, ..., 22-pq, 22qp.

Using the 44 networks, 6 level difference strings and absolute alignment probabilities we measured an error rate of 13% (Granum and Thomason, 1986). However, 6.3 of the 87% correctly type-classified had wrong orientation. In particular type 21 showed a symmetric pattern which confused the orientation in 37% of the cases, and types 16 and 19 in about 10% of the cases. Types 8, 9, 12 and 13 on the other hand never had best match in the opposite orientation.

We would expect that the use of the relative alignment probabilities will reduce these error rates by a few percent. In any case, doubling the number of networks has doubled the computational cost.

The networks are not absolutely symmetric representations of the band patterns, and we refer to the above determination of the orientation as an *asymmetric* method. By using not only 2 networks per class, or 2 alignments of each string (one direct and one reverse), *but both*, a *symmetric* estimation can be achieved. The string-to-class relationship for each string is then based on 4 alignments as illustrated below:

*Types of Alignment*

| String-Orientation | pq-Network | qp-Network |
|---|---|---|
| direct | **d-pq** | **d-qp** |
| reverse | **r-pq** | **r-qp** |

Results of alignments *d-pq* and *r-qp* then concern the same hypothesis and will *both* be based on either correct or incorrect orientation. Correspondingly, but opposite, for alignments *d-qp* and *r-pq*.

Using a combination of all 4 alignments for type-determination and then for given type their internal relationship for determination of orientation is expected to provide a rather robust estimate. This hypothesis has not been tested yet and will, if used, cause another doubling in the time for analysis.

## 5.3 Markov Networks and Centromere Position

Centromere information has initially been completely disregarded in the context of this method, as it is not essential to its basic operation. However, it may be beneficial to include that information, and recently the idea of introducing "Forced Landmarks" in the inference has been tested (Gregor, 1988). To force the centromere position to become a uniquely defined landmark in the final network all learning strings are separated into their respective p and q substrings. By inferring separate networks for p and q substrings, respectively, and then joining their networks together as one *FL-network* representing the whole chromosome, there is a possibility for specifying what position in an aligned string corresponds to the joint of the two networks. The hypothesis is then that this is the position of the centromere in the aligned string, if it belongs to the pattern class described by the network. Gregor (1988) tested this using a 4 level representation of C-group chromosome profiles of the database. On learning data the centromeres were correctly placed (±2 pixels) in 90.6% of the cases. Test results were 88.2% on average where the main reason for the difference stems from type 8, the test results of which dropped to 62% from the already poor learning results of 72%. The other results per type varied between 88.5% and 97.5%.

In recent experiments (Gregor *et al.*, 1988) we have repeated the procedure for *difference* symbols derived from both 4 and 6 level profiles. Test results were on average 94.7% and 94.3%, respectively, for these two data representations. With the difference symbol representations type 8 performed fairly close to the other types.

Hence results may be very dependent on specific structures of the individual chromosome types and on data representation, but the idea seems basically to be useful. We are looking forward to investigate the approach in more detail and for all chromosome types. In comparison Piper and Granum (1988) report 93.4% correct centromeres found for the C-group when testing automatic image analysis procedures using the same data base, and the same definition of correct position.

*Centromere finding from band pattern, C-group only*
*(given chromosome type and orientation)*

| Band representation | Error rate |
|---|---|
| 4 level strings | 11.8% |
| 4 level difference strings | 5.3% |
| 6 level difference strings | 5.7% |

## 5.4   Classification and Centromere Finding

The networks with forced landmarks can also be used for classification and Gregor (1988) has experimented with this. Firstly he showed that the generally smaller FL-networks performed slightly better in ordinary classification. Results were error rates of 21.3% (±14.7) and 23.6% (±17.4) respectively for group C. This was using the representation of 4 levels of density values. For comparison, the usage of 4 level difference strings showed only 15.9% (±12.4) errors (Gregor *et al.*, 1988). This indicates the importance of the representation scheme actually used (same raw profile data).

However, it is at least as interesting to consider the possibility of explicitly using the information about the forced landmarks. Each alignment of a string with a network is a test of the hypothesis that the string belongs to the class represented by the network. This hypothesis also implies a specific position of the centromere and the alignment provides an estimate of that position for the string in question. By combining the alignment probability with the associated estimate of the centromere position, a more robust and possibly improved classification can be obtained. Gregor (1988) has made a first attempt on this strategy using the difference between the estimated and the expected centromere position in a weighting factor applied to the alignment probability. In an experiment he showed that the error rate could hereby be reduced from the 21.3% (± 14.7) to 15.4% (±6.0) for the C-group chromosomes. More recent experiments (Gregor *et al.*, 1988) using 4 level difference strings and a slightly modified weighting showed 12.1% (±9.9) errors.

The improvements were mainly achieved by elimination of the typical confusion between type 6 and 10. Part of this problem was also eliminated by using *relative* alignment probabilities as discussed in Section 5.1. Using the relative alignment probabilities *also* in the above experiment reduced the error rate to 10.1% with a much smaller variation (±4.9).

*Within C-group classifications*
*orientation known; absolute alignment probabilities; 4 level string representation*

|  | density level strings | difference strings |
|---|---|---|
| Ordinary networks | 23.6% (± 17.4) | - |
| FL-networks | 21.3% (± 14.7) | 15.9%  (±12.4) |
| FL-networks & cent. est. | 15.4% (±  6.0) | 12.1%  (±9.9) |
| FL-networks & cent. est. (rel. align.) | - | 10.1%  (±4.9) |

## 5.5  Potential for full Karyotyping

On the basis of the experiments described above it may be interesting to speculate on a procedure for full karyotyping using these Markov methods. Input to the procedure would then be cell-wise sets of profiles with no more information than the profile and the chromosome length. The  procedure may include the following steps:

1. The length measures are cell-wise normalized and will thus on average provide a more narrow selection of networks to be tested per chromosome.
2. Profiles are processed to provide the appropriate string representation.
3. Strings are aligned with FL-networks as selected by the length test. Centromere position estimates are derived in connection with each alignment. Strings are aligned in direct and/or reverse orientation and/or with networks inferred for pq-and qp-orientation, respectively. That is, at least 2 alignments per string-type combination and possibly 4 alignments if it is found advantageous and worth the doubling of computing requirement.
4. The information from alignment and centromere estimation is combined and coordinated over the different orientations of string-to-network combinations.

5. Initial classifications are performed and then adjusted by a more or less sophisticated karyotyping procedure using cell context and expected class sizes.

If successful this procedure will generate information about class, pq-orientation and centromere position for each chromosome from cell-wise presented profiles. We have indications that motivate further investigations into this procedure. The eventual performance is uncertain, however. The best forecast we can make on results of analysing this data base is as follows: Isolated classification, 22 classes, known orientation gave about 8% error rate. Excluding information on orientation has been seen to cost about 3% more errors leaving us with 11% errors all together. Using strings and networks in both orientations may reduce errors by 1%. By adding centromere information we may gain another 3% in performance, so that expected performance on isolated classification now corresponds to 7% error rate. Also introducing cell-wise normalization of length and a karyotyping strategy could possibly correct half the errors leaving us eventually with, say 4% error rate, or about 2 errors per cell on these data.

Such figures compare with previous results on the same data base. Using the global WDD-features Granum (1982) obtained 5.1% and 2.5% error rates for isolated classification and karyotyping, respectively, on the very same profile data. These results, however, assumed correct pq-orientation and centromere position provided by the image preprocessing. In case of the Markov models, the classification and karyotyping is carried out without this a priori (and possibly uncertain) information. On the contrary, that information is an output of this alternative approach. A more realistic reference on automatic analysis of these data is provided by Piper and Granum (1988), who used the Edinburgh chromosome analysis system. They karyotyped with 5.8% errors, using automatic centromere and orientation finding and having bent chromosomes included.

Still, an estimate of about 4% error rate on this data base of "polished" data does not tell final performance on data acquired and analysed in a routine clinical context, and the structural approach may well for natural reasons be more sensitive to bad quality banding.

## 6.  Computational Cost

The measures of eventual interest for evaluation of the computational cost of using this approach are in particular memory requirements and time taken for analysis of one cell. Memory requirements are up to the order of 1 Mbyte and hence acceptable by today´s technology. Execution time is related to the number of alignments to be computed, and this again depends on the details of the approach. Typical figures are listed for the different variants of the approach discussed above:

| PQ-Orientation | Pre-Class. | Symmetry | Appr. No. of Align.s/Cell |
|---|---|---|---|
| Known | Denver Groups | (Asymmetric) | 200 alignments |
| Known | Length Test | (Asymmetric) | 150 alignments |
| Unknown | Length Test | Asymmetric | 300 alignments |
| Unknown | Length Test | Symmetric | 600 alignments |

The time per alignment depends on the length of the chromosome and the size of the network. On a VAX 8200 computer using the original flexible experimental code (highly inefficient) the following typical execution times per alignment were observed:

| Type 2 | 80 symbols per string | 400-600 states per network | 20 sec. |
| Types 8-11 | 50 symbols per string | 250-460 states per network | 9 sec. |
| Types 21-22 | 25 symbols per string | 110-225 states per network | 2 sec. |

On average over all the chromosomes of a cell, 10 sec. per alignment is considered representative for this specific implementation. However, the power of the computer as well as optimisation of the code for the specific purpose has a significant impact on these timings. Cole (1988) used a VAX 8700 and implemented the dynamic programming for computation of alignment only (saving no information for the network update of the inference mode etc.) and he measured .32 sec. on average per alignment - a factor of 30 faster than the original code. The range of time of analysis per cell for these implementations is then:

| | ------------- *Implementation* --------------- | |
| *No. of alignments* | *Original/VAX 8200* | *Cole/VAX 8700* |
| 150 | 25 min. | 0.8 min. |
| 300 | 50 min. | 1.6 min. |
| 600 | 100 min. | 3.2 min. |

Essentially there are no particularly discouraging elements in the computational requirements. By the time the approach might be ready for routine use, available standard hardware will be able to execute appropriate implementations of the analysis sufficiently fast.

The time taken to infer a network of average size using 100 learning samples is about 10 min. using the original implementation. Hence 24 networks can be inferred in less than 4 hours, and no prohibitive requirements for practical application appear here either.

# 7.   Miscellaneous Extras

## 7.1   Pruning of Networks

Inferred networks contain a large number of nodes and arcs of very low frequency counts corresponding to very rarely observed band pattern variations. Nielsen (1988) has studied a method for systematically eliminating (pruning) such low frequency elements from the networks and evaluated the consequences on classification performance etc. He found that a modest pruning of 10% halved the number of nodes and arcs without noticeably degrading performance. One pruning operation included the identification of the least likely path through the network and reducing all counts along this path by one. Nodes and arcs having frequency counts of one will typically be encountered on these paths, and after the operation they are empty and can be removed from the network. After an operation the network will appear as if the learning set of samples had one string less than actually used. After 10% pruning of a network inferred from 100 samples the resulting network will look as if only 90 samples had been used. It will have the same core structure of high frequency elements and hardly any fringes of low frequency.

The computational cost is proportional to the number of nodes and arcs of the networks. The use of pruned networks would then possibly more than halve the time for analysis. This gain in speed has not been include in the previous discussions, but the option will be re-evaluated when the general approach has matured further.

### 7.2 Analysis of Band Pattern Normality

Nielsen (1988) has also made a pilot study of the possible use of the Markov models for analysis of the normality of a band pattern given its chromosome type. Although a potential is intuitively there, Nielsen concluded that it is not obvious how to bring it about. Activities in this respect are awaiting more insight, new ideas and/or a very specific task.

## 8.  Summary and Conclusions

A structural approach to analysis and classification of chromosome band patterns has been presented. The presentation included discussions of (i) band pattern representation, (ii) a syntactic method for analysis and classification, which includes facilities for automatic inference (learning) of models to be used, and (iii) evaluation of the method on the basis of practical experiments.

(i)     The discussion on band pattern representation focused on local and sequential properties with particular emphasis on conserving all detectable transitions between bands. For the syntactic analysis a *string* representation was produced from processed profiles of 4 - 6 "density levels". It has been found that strings representing the transitions ("density level" differences) give better classification results than direct encoding of the "density levels".

(ii)    The syntactic method includes automatic inference of the models (Markov networks) used to represent the sequential structure of the classes. Both inference and classification rely on alignment of sample strings with network models, and optimal alignments are computed using dynamic programming. Each network is modelling chromosome profiles (strings) of one class and all in same pq-orientation, and centromere position need not be considered. However, the method may be applied such that the orientation may be estimated rather than known a priori, and the inference process may be carried out with a forced landmark such that the centromere position can be estimated in connection with the classification procedure. Further, by considering consistency between class estimates and estimates of centromere position, higher classification rates can be achieved.

(iii)   Experimental evidence of the basic potential of the method is provided using a rather "clean" data base of profiles of straight chromosomes. Results obtained compared to those of other approaches when analysing the same data base. The major result so far is an error rate of 7.7% in classifying the 22 autosomes knowing the pq-orientation but not using the centromere. (Within the C-group only 12.4%.). Using networks with a forced landmark at the centromere, the position of C-group centromeres could be estimated with only 5.3% errors given the chromosome type. Classifying the C-group, estimating centromeres and considering the consistency of estimated class and centromere position gave 10.1% classification errors.

From the results it is concluded that this structural approach has a potential for practical use in chromosome analysis. Combining our experience with these methods and with chromosome analysis in general we predict that a full karyotyping procedure based on this approach may perform as well on the data base used, as have other methods using global features.

The final conclusion is not necessarily that yet another competing approach has appeared on the scene. Rather, one should notice that the type of information exploited by this approach is sequential structure of local band pattern features, which presumably has a low correlation with the information used in other approaches. Hence the possible importance of this contribution may appear as a supplement to other methods based on global information.

*Acknowledgements.* Since this research was initiated in 1982 partial support has been received from the Burroughs Welcome Fund, the UK Medical Research Council, the Danish Technical Research Council, the NATO Office of Scientific Affairs, The Danish Council for Research Policy and Planning, the US National Institute of General Medical Sciences and the CEC Medical and Health Research Programme. Lisbeth Frederiksen helped in preparation of the manuscript.

# References

1    An International System for Human Cytogenetic Nomenclature (1985). Report of the standing committee on human cytogenetic nomenclature. S. Karger Publishers, Basel (1985)

2    Casperson T, Lomakka G, Møller A. Computerized chromosome identification by aid of the Quinacrine mustard fluorescence technique. Heredita, 67, pp. 103-109 (1971)

3    Cole GS. Constrained Dynamic Programming Inference of Markov Networks From Finite Sets of Sample Strings. M.Sc. Thesis, Computer Science Department, University of Tennessee, Knoxville (June 1988)

4    Gonzalez RC, Thomason MG. Syntactic Pattern Recognition: An Introduction. Addison-Wesley, Reading, MA (1978)

5    Granlund GH. The use of distribution functions to describe integrated density profiles. Journal of Theoretical Biology 40, pp.573-589 (1973)

6    Granum E, Lundsteen C. Progress on Band Pattern Derivation from Digitized Chromosome Images. Proc. of IV Nordic Meeting on Medical and Biological Engineering (IFMBE), Technical University of Denmark, Lyngby (1977)

7    Granum E. Pattern recognition aspects of chromosome analysis - Computerized and visual interpretation of banded human chromosomes. Ph.D. Thesis, Laboratory of Electronics, Technical University of Denmark, Lyngby (1980)

8    Granum E. Application of statistical and syntactical methods of analysis and classification to chromosome data. In: J. Kittler, K.S. Fu and L.F. Pau (eds), Pattern Recognition Theory and Applications, NATO ASI (Oxford, 1981), Reidel, Dordrecht (1982)

9    Granum E, Thomason MG. Classifying Chromosome Band Patterns using Automatically inferred Markov Networks. 8th European Chromosome Analysis Workshop, Berlin (1986)

10 Granum E, Thomason MG. Chromosome Analysis and Automatically Inferred Markov Networks I. Classification of Band Pattern Structures. To be submitted to Computers in Biology and Medicine (1988)

11 Gregor J. Inference of Markov Networks with Forced Landmarks. M.Sc. Thesis, Institute of Electronics Systems, University of Aalborg (June 1988)

12 Gregor J, Granum E, Thomason MG. Chromosome Analysis and Automatically Inferred Markov Networks II. Finding C-group Centromeres Using Band Pattern Information (In preparation, 1988)

13 Kramer HP, Bruckner JB. Iterations of a Non-linear Transform for Enhancement of Digital Images. Pattern Recognition 7, pp. 53-58 (1975)

14 Ledley RS, Ing PS, Lubs HA. Human chromosome classification using discriminant analysis and Baysian probability. Comput. Biol. Med. 10, pp.209-218 (1980)

15 Lundsteen C, Granum E. Visual classification of banded human chromosomes II. Classification and karyotyping of integrated density profiles. Ann. Hum. Genet. 40, pp. 431-442 (1977)

16 Lundsteen C, Granum E. Visual Classification of Idealized Profiles (1978). Unpublished

17 Lundsteen C, Granum E. Description of chromosome banding patterns by band transition sequences. Clinical Genetics 15, pp. 418-429 (1979a)

18 Lundsteen C, Granum E. Visual classifcation of banded human chromosomes III. Classification and karyotyping of density profiles described by band transition sequences, Clinical Genetics 15, pp.430-439 (1979b)

19 Lundsteen C, Phillip J, Granum E. Quantitative analysis of 6985 digitized trypsin G-banded human metaphase chromosomes. Clinical Genetics 19, pp. 355-370 (1980)

20 Lundsteen C, Gerdes T, Granum E, Phillip J, Phillip K. Automatic chromosome analysis II. Karyotyping of banded human chromosomes using band transition sequences, Clinical Genetics 40, pp. 26-36 (1981)

21 Nielsen TF. Aspects of Dynamically Programmed Inference of Markov Networks. M.Sc. Thesis, Institute of Electronics Systems, University of Aalborg (February 1988)

22 Piper J, Granum E. On fully automatic feature measurement for banded chromosome classification. Submitted to Cytometry (1988)

23 Piper J, Granum E, Rutovitz D, Ruttledge H. Automation of chromosome analysis. Signal Processing 2, No. 3, pp. 203-221 (1980)

24 Ten Kate TK, Groen FCA, Young IT, van der Ploeg M, Pearson PL, Gerdes T, Lundsteen C. The Delft classification technique on some difficult chromosome classes. Proceedings of the Vth European Chromosome Analysis Workshop, Heidelberg (1983)

25 Ten Kate TK. Design and implementation of an interactive karyotyping program in C on a Vicom digital image processor. M.Sc. Thesis (in Dutch), Department of Applied Physics, Delft University of Technology , Delft (1985)

26 Thomason MG, Granum E. Dynamic programming inference of Markov networks from finite sets of sample strings. IEEE Transactions on Pattern Analysis and Machine Intelligence, PAMI-8, No. 4, pp. 491-501 (1986)

27 Thomason MG, Granum E, Blake R. Experiments in dynamic programming inference of Markov networks with strings representing speech data. Pattern Recognition 19, No. 5, pp. 342-35 (1986)

28 Thomason MG, Granum E. Sequential inference of Markov networks by dynamic programming for structural pattern recognition. Pattern Recognition Letters 5, pp. 31-39 (1987)

29 Thomason MG. Structural methods in pattern analysis. In: P.A. Devijver and J. Kittler (eds), Pattern Recognition Theory and Application, NATO ASI (Spa, 1987), Springer-Verlag, Berlin (1987)

30 Visser RT. Classification of banded chromosomes using a priori cytologic knowledge. M.Sc. Thesis (in Dutch), Department of Applied Physics, Delft University of Technology, Delft (1981)

31 Zimmerman SO, Johnston DA, Arright FE, Rupp ME. Automated homologue matching of human G-banded chromosomes. Comput. Biol. Med. 16, No. 3, pp. 223-233 (1986)

# Density Profiles in Human Chromosome Analysis

A. Forabosco, P. Battaglia, R.Bolzani

**Summary**

Our experience with a general purpose image analyzer oriented to the automated analysis of chromosomes is described. The system allowed us to carry high resolution studies on the banding pattern of R-banded human chromosomes at different degrees of condensation. Density profiles were extracted by a simple method through the definition of the median axis of the chromosomes and then calculating the average density for each point of the segments that cross the longitudinal axia. These density profiles can be used to analyse the banding pattern of homologous chromosomes, metaphase or prometaphase, or to create standard profiles for automatic chromosome classification.

## 1. Introduction

Cytogenetic analysis is well suited for diagnosing chromosome abnormalities and is currently an irreplaceable instrument in preventive medicine programs fighting congenital malformations. Moreover, it is an important means for diagnosing neoplastic forms and monitoring the biological effects of environmental mutagens. These wide possibilities have led to an increasing request for cytogenetic testing, which however cannot be readily met owing to the lengthy analysis procedure. Two solutions have been proposed to reduce the time required to perform the analysis, flow cytophotometry [1] and automatic image analysis. Activity in this latter field, which began in the early seventies, now includes several commercially available instruments consisting of image processors and a group of computer programs used for the routine analysis of metaphase chromosome preparations [2,3].

Techniques have recently been developed to obtain prometaphase preparations, which notably increase the degree of resolution of the chromosome. High resolution testing with prometaphase chromosomes is important owing to the recent discovery of microcytogenetic chromosome pathologies such as Retinoblastoma or Wilm's aniridia-tremor syndrome. Duchenne's muscular dystrophy has also been shown to result from a submicroscopic deletion [4].

Our interest in studying metaphase and prometaphase chromosomes has required the development of a procedure for high resolution analysis of individual chromosomes starting from their photographic image.

**Automation of Cytogenetics**   Editors: C. Lundsteen J. Piper
© Springer-Verlag Berlin Heidelberg New York 1989

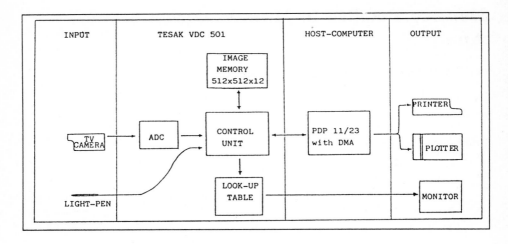

**Fig. 1.** Configuration of the system

## 2. Hardware specification

We utilized a general purpose system that did not enable us to realize a complete automated chromosome analysis. Nevertheless its ability to elaborate images interactively makes this system suitable for partial analysis. The hardware configuration of our system, shown in figure 1, consists of four main blocks, input devices, image processor (TESAK VDC 501), host computer, and output devices. The VDC is linked, through a parallel interface, with the host computer which supervises the whole system and drives the VDC to execute operations on the image memory. The image memory is organized as 12 planes of $512 \times 512$ pixels. Each column of that three-dimensional structure represents a 12 bit value for the selected pixel information; the first 8 bits are reserved for the intensity and the remaining 4 bits are available for overlaying. Therefore the intensity can be represented with $2^8 = 256$ density levels.

The analog input image coming from a black and white TV camera is digitized by an analog-to-digital converter (ADC) and written on the first 8 planes of the image memory, which code the pixel density levels.

The display shows the acquired image frame by frame through a look-up table (LKT) coding. In figure 2 the LKT is schematically shown as a memory of 4096 locations of 16 bits. Each 16 bits location contains the three basic color contributions and common offset. There is a one-to-one correspondence between the $2^{12}$ possible pixel values and the LKT addresses, so each pixel value addresses one and only one LKT location which might have been previously programmed to obtain the desired color or grey tone. In this way 4096 different colors and 256 grey tones can be obtained.

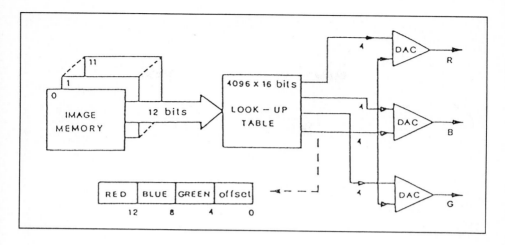

**Fig. 2.** Look-up table coding

## 3. Software characteristics

The images are obtained by microphotography or by means of a microscope using a black and white TV camera. They are converted by the ADC and finally displayed on the monitor. The operator adjusts frame and focus directly on the digitized video frame. After the set-up operation the image is fixed. When the picture is judged to be as good as it can be, the operator sends it to the computer using the key-board and there it is stored on the mass memory. Our method, to obtain maximum image definition, is to frame a single chromosome on the screen, with a spatial resolution of 10 pixel/$\mu$m, and then to determine its median axis. We found that due to the architecture of the system, the automatic process was much slower than interactive techniques, which have enabled us to deal with particular circumstances, in which a corrective intervention would have been required in any event, with much greater flexibility. The slow speed of the system is due to the fact that the image memory is not part of the host computer, and therefore must go through the VDC (image analyzer) control unit to access the values of the individual pixels.

The interactive technique requires the operator to use the light-pen to identify the ends of the segments forming the central axis, and these are then used by the computer to construct the entire polygon. The number of segments of the polygon depends on the curvature of the chromosome (figure 3). To calculate the density curves along the length of the chromosome, the straightened image must first be re-constructed, and then the average density can be calculated for each point of the segment that crosses the longitudinal axis.

Straightening is achieved by dividing the image into perpendicular strips at the median axis and then re-setting the strips on a straight axis. The average densities of the various strips form the density curve of the chromosome [5].

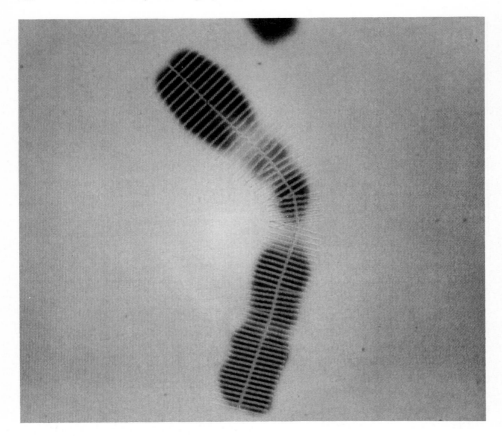

**Fig. 3.** Longitudinal axis

Strictly speaking, it is not necessary to re-construct the straightened image since it is sufficient to obtain the average density for each strip. In fact our program shows the whole chromosome image only if requested by the operator and by default provides the chromosome density curves. The density curve is used to identify the various bands on the chromosome. This is achieved by filtering the curve with a digital band-pass filter. The filter is realized by using a two stage moving averages function. The filter removes the high frequency component. Since the moving average function is a low pass filter, the second step provides the medium frequency component subtracting the very low frequency component from the first stage output. The resulting formula is:

$$D_i = \frac{1}{2n+1} \sum_{k=-n}^{n} D_{i+k} - \frac{1}{2m+1} \sum_{j=-m}^{m} D_{i+j},$$

with $n = 7$ and $m = 17$. To eliminate phase inversion each stage is applied

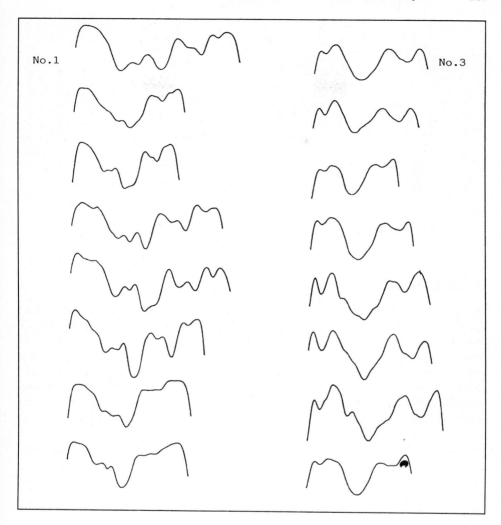

**Fig. 4.** Number 1 and 3 RHG chromosome profiles

twice. The resulting bandwidth of the filter ranges from 21.7 to 52.6 pixels. Finally, the maximum values, representing the position of the identified bands, are found from the filtered curve.

## 4. Applications

The methodology described has been used on metaphase or prometaphase chromosomes from peripheral blood lymphocyte cultures with RHG or RHTBG banding pattern.

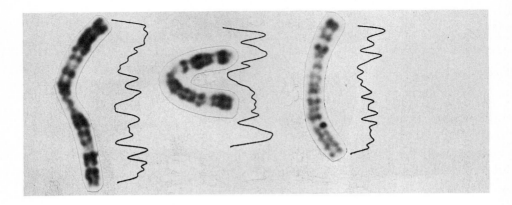

Fig. 5. RHTBG banded prometaphase normal chromosomes (1, 3, 4)

## 4.1. Chromosome analysis

The density profiles of metaphase and prometaphase chromosomes are used to identify the bands as density peaks, the presence and relative position of which can be utilized to compare homologous chromosomes (figure 4). The automation of prometaphase chromosome analysis presents certain problems which makes it completely different from automatic karyotyping of metaphases. These chromosomes are more elongated and are often notably bent and overlapping. Furthermore, the greater definition of chromosomal bands requires high resolution digital conversion of the image. In fact, the spatial frequency pattern along a prometaphase chromosome is notably greater than that in metaphase (figure 5). Naturally this type of testing is not a routine diagnostic procedure and does not require the study of the entire metaphase but only of specific chromosomes. An example of automatic analysis of banding pattern between metaphase and prometaphase homologous chromosomes with a structural abnormality is shown in figure 6.

## 4.2. Chromosome classification

Automatic chromosome classification is still a problem under discussion. Granum [6,7] and Piper [8] have suggested and tested statistical and syntactical methods of analysis and classification. We intended to verify the applicability of statistical techniques based on the correlation coefficient as a classification function. We used density profiles to create standard profiles of metaphase chromosomes which can be used for automatic chromosome classification. An experiment was carried out to verify the efficiency of a classification procedure based on comparing chromosome density profiles and standard density profiles using the correlation coefficient. For each chromosome type the average length

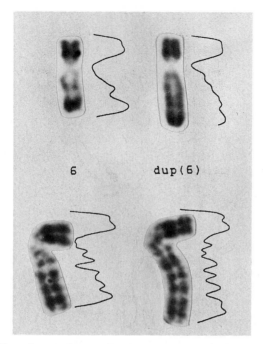

**Fig. 6.** Density profiles of metaphase and prometaphase chromosomes no. 6. On the right the chromosomes with structural abnormality

was found and used to normalize the length of 38 homologous chromosome profiles. Then the standard profile of each chromosome of the human karyotype was obtained from a point by point average of the normalized profiles (figure 7). The standard profiles thus evaluated were used to classify 20 metaphases. Each chromosome to be classified was compared to all standard profiles whose length differed by less than 30%. The index of comparison was the coefficient of correlation between the two curves set to the same length. The chromosome being tested was associated with the standard profile having the largest coefficient of correlation. The resulting classification matrix is presented in figure 8.

# 5. Conclusions

The classification method utilized only uses the information retained in the density profile of the chromosome to be analyzed. This makes this method insensitive to errors due to the calculation of the centromere position, and avoids

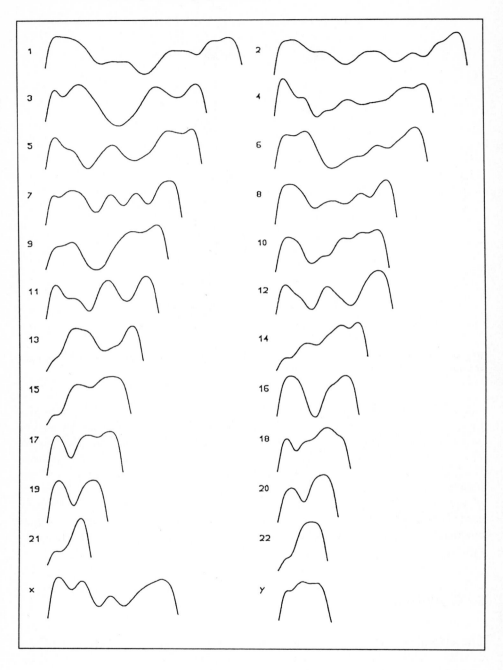

**Fig. 7.** Standard profiles

operator assignement

| | 1 | 2 | 3 | 4 | 5 | 6 | 7 | 8 | 9 | 10 | 11 | 12 | X | 13 | 14 | 15 | 16 | 17 | 18 | 19 | 20 | 21 | 22 | Y | |
|---|---|---|---|---|---|---|---|---|---|---|---|---|---|---|---|---|---|---|---|---|---|---|---|---|---|
| 1 | 34 | 2 | 1 | | | 1 | | | | | | | | | | | | | | | | | | | |
| 2 | 2 | 32 | | | | 1 | 1 | | | | | | | | | | | | | | | | | | a |
| 3 | | | 35 | | | 1 | | | 1 | | | | | | | | | | | | | | | | u |
| 4 | 1 | 1 | | 31 | 5 | | | 1 | | | | | | | | | | | | | | | | | t |
| 5 | 1 | | | 3 | 30 | | | | | | | | 2 | | | | | | | | | | | | o |
| 6 | | | | 2 | 3 | 31 | 2 | | 1 | | 1 | | | | | | | | | | | | | | m |
| 7 | | 2 | | | | 1 | 36 | 4 | 2 | 2 | | | | 1 | 1 | | | | | | | | | | a |
| 8 | | | | | | 1 | 1 | 30 | 4 | | | | | 1 | 1 | 1 | | | | | | | | | t |
| 9 | | | | | | 1 | | | 29 | 4 | | | | | | 1 | | | | | | | | | i |
| 10 | | | | | | | | 2 | 1 | 24 | | | | 1 | | | | 1 | | | | | | | c |
| 11 | | | | | | | | | 1 | 1 | 37 | | | | | | | 1 | | | | | | | |
| 12 | | | 1 | 1 | | | | 1 | | | | 34 | | 1 | | | | | | | | | | | a |
| X | | 1 | | | | | | | 2 | | | | 25 | | | | | | | | | | | | s |
| 13 | | | | | | | | 1 | | 1 | | 1 | | 35 | 2 | | | | | | | | | | s |
| 14 | | | | | | | | | | | | | | | 28 | 7 | | | | | | | | | |
| 15 | | | | | | | | | | | | | | | 3 | 27 | 1 | | | | | | | | |
| 16 | | 1 | | | | | | | 1 | | | | | | | 1 | 30 | 1 | | 2 | | | | | |
| 17 | | | | | | | | | 1 | | 1 | | | | | | | 33 | 6 | | | | | | |
| 18 | | | | | | | | | | | | | | 3 | | | 2 | | 29 | 2 | | | | 2 | |
| 19 | | | | | | | | | | | | | | | | 5 | | | | 29 | 5 | | | 4 | |
| 20 | | | | | | | | | | | | | | | | | 2 | 2 | 1 | 5 | 30 | 1 | | 1 | |
| 21 | | | | | | | | | | | | | | | | | | | | | 3 | 34 | 8 | | |
| 22 | | | | | | | | 1 | | | | | | | | | | 1 | | | | 2 | 29 | | |
| Y | | | | | | | | | | | | | | | | | | 1 | | | 1 | 1 | 1 | 3 | |
| tot | 38 | 38 | 38 | 38 | 38 | 38 | 38 | 38 | 38 | 38 | 38 | 38 | 28 | 38 | 38 | 38 | 38 | 38 | 38 | 38 | 38 | 38 | 38 | 10 | |

(right margin, reading top to bottom: automatic assignment)

**Fig. 8.** Classification matrix

the need for interactive correction of it. On the other hand the correlation coefficient alone is not sufficient to produce a correct classification of all the chromosomes. This is due to the fact that the chromosome density profiles vary depending on the degree of condensation. A solution could be to create different classes of standard chromosome profiles of different degrees of condensation and to choose the class of standard profiles to be compared with the chromosomes to be classified on the basis of the metaphase length.

We are working to verify the validity for practical use of density profiles in metaphase and prometaphase chromosome studies.

*Acknowledgements.* This work was partially supported by the Commission of the European Communities within the frame of its Medical and Health Research Programme by project no. II.1.1/13, and by the Consiglio Nazionale delle Ricerche-Progetto Finalizzato Tecnologie Biomediche e Sanitarie, project no. 87.007.45.57.

# References

1    Young BD, Ferguson-Smith MA, Sillar R, Boyd E. High resolution analysis of human peripheral lymphocyte chromosomes by flow cytometry. Proc.Natl.Acad.Sci. 78:7727-7731 (1981)

2    Lundsteen C. Automated karyotyping systems. Report on Eur. Work Group on Autom. Chrom. Anal. (Berlin, 1986)

3    Piper J, Granum E, Rutovitz D, Ruttledge H. Automation of chromosome analysis. Signal Proc. 2:203-221 (1980)

4    Dallapiccola B, Forabosco A. Human microcytogenetics. Acta Med. Auxol. 19:5-33 (1987)

5    Piper J. Interactive image enhancement and analysis of prometaphase chromosomes and their band patterns. Anal. Quant. Cytol. 4:233-240 (1982)

6    Granum E, Gerdes T, Lundsteen C. Simple weighted density distributions, WDDs for discrimination between G-banded chromosomes. Proc.IVth European Chromosome Analysis Workshop (Edinburgh, 1981)

7    Granum E. Application of statistical and syntactical methods of analysis and classi-fication to chromosome data. In: Kittler J, Fu KS, Pa LF (Eds) NATO ASI series no. C.81: Pattern Recognition theory and Applications (D. Reidel, Dordrecht, 1982) pp.373-398

8    Piper J. Finding chromosome centromeres using boundary and density information. In: Simon J-C, Haralick RM (Eds) NATO ASI series no. C.77: Digital Image Processing (D. Reidel, Dordrecht, 1981) pp.511-518

# Cytogenetic Analysis by Automatic Multiple Cell Karyotyping

Claes Lundsteen, Tommy Gerdes, Jan Maahr

**Summary**

To increase the efficiency of the Magiscan system for automatic metaphase finding and karyotyping we propose a procedure for multiple cell karyotyping. While a metaphase is counted by the operator with a lightpen the previous counted metaphase will automatically be karyotyped. From 10-15 karyotyped metaphases a composite karyotype will be printed out to search for structural abnormalities. Due to segmentation and classification errors on the average only 53% of the chromosome pairs will automatically be correctly classified in the composite karyotype. However, this is supposed to be sufficient for practical purposes. The speed of the analysis may be increased by a factor of two to approximately 15 minutes per case. As a spin off of the multiple cell approach a procedure for automatic detection of minor structural abnormalities and a procedure for automatic quality assessment have been designed.

## 1. Introduction

During the past few years we have witnessed impressive advances in the development of commercially available systems for automated cytogenetic analysis (Lundsteen and Martin, 1989). Some systems can perform both automatic metaphase finding and karyotyping, while others only perform karyotyping.

We have been using the Magiscan chromosome system for clinical cytogenetic analyses since 1983. Extensive clinical trials have demonstrated that the karyotyping-only system produces one karyotype in approximately 7 minutes, which is 4 - 5 times faster than by manual means. With the metaphase finding + karyotyping system one "case", comprising 10 counts, 3 "eye ball" karyotypes (i.e. visual search for structural abnormalities in the metaphase) and 1 "machine" produced karyotype, is carried out in approximately 35 minutes, which is about half the time required for a corresponding manual analysis (Lundsteen et al., 1987).

Chromosome systems are expensive and as the karyotyping-only system is not only faster but also cheaper than the metaphase finding + karyotyping system, the former is apparently most cost effective (Lundsteen and Martin, 1989). The performance of the latter system might, however, be improved -

**Automation of Cytogenetics**   Editors: C. Lundsteen J. Piper
© Springer-Verlag Berlin Heidelberg New York 1989

without significant increase of the cost - if the way the analysis is carried out might be changed.

We have therefore looked for ways to increase the speed and to reduce operator interactions of the metaphase finding + karyotyping system by examining where in the analysis operator time and interactions might be saved. Based on time recordings of the individual steps of the analysis we found that a considerable part of the operator time is spent during "eye ball" karyotyping and "machine" karyotyping. "Eye ball" karyotyping takes time because each chromosome pair has to be located, matched and examined for structural abnormalities. "Machine" karyotyping takes time because the operator has to assist the system by separating touching and overlapped chromosomes and to correct misclassified chromosomes with the lightpen.

In our multiple cell approach we avoid "eyeball" karyotyping and ignore errors in the karyotype due to overlapped, poorly banded or distorted chromosomes, because these chromosomes are not very useful for the detection of structural abnormalities anyway. Therefore, it may not be worth spending time on lightpen interactions correcting these chromosomes. Instead more metaphases are karyotyped automatically and combined in a composite karyotype, where all karyotyped metaphases are represented by a column of 23 chromosome pairs and where all columns are placed next to each other, so that chromosome pairs of the same type are lined up. The composite karyotype is printed out and forms an excellent basis for the search for structural abnormalities.

This multiple cell approach offers two additional advantages both of which are due to the fact that a number (e.g. 10) of metaphases are karyotyped. The measurements obtained are used for automatic detection of minor structural abnormalities and the same or related measurements may form the basis for an automatic assesment of slide quality.

The purpose of this paper is to describe the multiple cell approach and to present some preliminary performance data.

# 2. Material and methods

## 2.1. System configuration and operation

The Magiscan chromosome system has previously been described in detail (Lundsteen et al., 1987). It consists of a Magiscan image processor (Joyce-Loebl) connected to a microscope with a TV-camera and a motorized eight slide scanning stage. Metaphases to be analysed are displayed on a monitor and the operator can assist the system in the analysis by using a lightpen to draw lines and mark objects on the screen and to give orders by a keyboard. The system is connected to a hard copy printer (Honeywell VGR 5000) which can produce hard copies of the generated karyograms.

The traditional way of using the system is as follows: Automated metaphase finding is performed over night. The scanning stage is loaded with up to 16 slides. For each step of the stage the field of view is analysed for the presence of metaphases, the coordinates of which are stored on disc. The located metaphases are automatically ranked according to their suitability for analysis.

Chromosome counting and "eye ball" karyotyping are done without automation simply by using the lightpen to mark the chromosomes which are visually counted and analysed respectively.

"Machine" karyotyping is initiated by an automatic segmentation of the metaphase image into isolated chromosomes. The operator uses the lightpen to separate touching and overlapped chromsomes which are incorrectly processed by the software. Axes, centromeres and banding patterns are automatically determined, the chromosomes are classified and the karyotype is displayed on the moniter. Misclassified chromosomes are corrected with the lightpen and the approved karyotype printed out.

In a previous study the time spent on individual steps of the analysis was recorded (Lundsteen et al., 1987). The time per metaphase count was 52 seconds, per "eye ball" karyotype 3 minutes, per "machine" karyotype 7 minutes and the "relocation" time (i.e. time "between" metaphases, waiting for the stage to move, for adjusting metaphase position, and focus) was 10 minutes per case.

## 2.2. The multiple cell approach

**2.2.1. Composite karyotyping.** It was realised that without major hardware modifications it would not be possible to obtain a significant reduction of the "relocation" time, which constituted a considerable part of the total time per case.

It was further decided not to change the manual counting of the chromosomes with the lightpen because: 1) manual counting is fast and reliable, 2) correct automatic counting is not yet achievable, especially if mosaicism is taken into account (Piper and Lundsteen, 1987), 3) manual counting gives the operator an extra possibility to make sure that all chromosomes belonging to the metaphase are within the field of view, and 4) the chromosome count is considered the most important part of the analysis of most constitutional abnormalities.

The time spent on three "eye ball" karyotypes + one "machine" karyotype was approximately half the total time for analysing one case. This time could be considerably reduced by omitting "eye-ball" karyotyping and by avoiding operator interaction during "machine" karyotyping.

Based on these considerations the following strategy for the analysis is proposed:

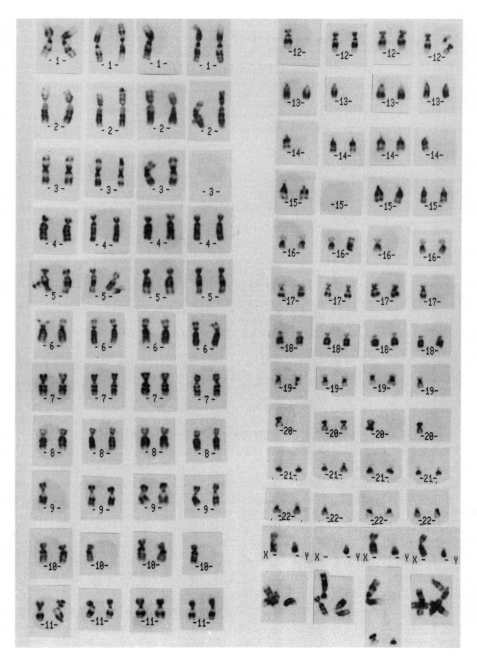

**Fig. 1.** Example of composite karyotype produced by cutting the individual chromosome pairs out of 4 hard copy karyotypes obtained from one routine analysis. Karyotyping was done without operator interaction. The resulting errors, however, should not hamper the search for structural abnormalities.

The operator relocates metaphases from the ranking list in the usual way. If the metaphase is suitable for counting he/she counts it in the usual way with the lightpen. If the metaphase is suitable for karyotyping, the digitized image of the metaphase is automatically karyotyped and the karyotype image stored in the image memory. While the automatic karyotyping goes on the operator continues the analysis by relocating the next cell in the ranking list which is counted and if suitable also karyotyped without operator interaction as described above. This procedure continues until an appropriate number of metaphases are counted (e.g. 10 - 15) and an appropriate number of karyotypes are stored in the image memory (e.g. 10). At this stage a composite karyotype is printed out by the hard copy printer (Fig. 1). The chromosomes are arranged so that comparison of chromosome pairs from the individual metaphases is facilitated. Because the karyotyping was done without operator interaction overlapped chromosomes, grossly abnormal chromosomes and other chromosomes which cannot be classified will be missing in the composite karyotype, but as seen in Fig. 1 they will be shown underneath the classified chromosomes. The composite karyotype forms the basis for the search for structural abnormalities. Grossly abnormal chromosomes will be detected, because they are misclassified and therefore no complete pair of homologues will be present in the composite karyotype. Minor structural abnormalities may be detected by visual comparison of the homologues of each of the 23 pairs or automatically as described below.

**2.2.2. Automatic detection of minor structural abnormalities.** In the conventional cytogenetic approach to detect small structural abnormalities like single band deletions the cytogeneticist compares the banding patterns from selected well banded homologous chromosomes, and if a difference in the banding pattern between the homologues is found in one cell it may be confirmed by looking at other cells. In most cells some chromosomes will be either overlapped or the banding patterns are poor or distorted. Therefore a number of cells will have to be examined before all pairs have been checked.

Differences between homologous chromosomes due to structural abnormalities may be reflected in selected appropriate measurements. It might therefore be possible to design a procedure for automatic detection of differences which would indicate the presence of a structural abnormality which eventually could be verified by the cytogeneticist. Such a procedure would be helpful in the diagnostic work, if it could draw the attention of the cytogeneticist to small aberrations that he/she otherwise would overlook and it might also be possible to quantify the aberrations in terms of amount of missing or additional chromosome material or a change in the centromeric index.

One possibility would be to determine the difference between measurements of homologous chromosomes. A difference averaged over a number of cells exceeding a certain predefined limit could indicate the existence of a structural abnormality.

A second approach which we attempted a few years ago was based on the observation that if one of two homologous chromosomes was abnormal its measurements deviated more from the mean measurement of that chromosome type than the normal homologue (Lundsteen et al, 1983). We used the same measurements (features) that we are currently using in our banded classifier (Lundsteen et al. 1986). For each chromosome pair we considered only the homologue which measurements showed the longest distance to the mean measurements of normal chromosomes. If this distance exceeded a certain arbitrary limit when averaged over a number of cells, the chromosome was considered possibly abnormal. The procedure was tested retrospectively on 25 minor structural aberrations. 19 were detected and 6 were missed when the arbitrary limit was set so that the false positive rate was one pair of homologues in every three patients (Lundsteen et al. 1983). The procedure was never implemented and used for clinical purposes, partly because poorly banded and distorted chromosomes tented to blur the measurements.

We have now designed a third way for automatic detection of structural abnormalities. A structural abnormality is best detected in an elongated well banded chromosome by comparing it with its well banded normal homologue. Therefore the "nicest" pair of homologues of each chromosome type is selected automatically from the composite karyotype on the basis of its measurements. Selection criteria are the number of bands and the smallest average distance from the pair of homologues to the mean measurements of the normal chromosomes. The closer to the mean the "more normal" are the chromosomes supposed to be, and opposite if one of the homologues of the "nicest" pair deviates to a certain degree from the normal chromosomes as well as from the normal homologue it may be an abnormal chromosome.

The procedure described above has not yet been tested. However, based on our experiences with our previous procedure we believe that it may work in practice.

**2.2.3. Automatic assesment of the quality of cytogenetic preparations.** It is of great importance for cytogenetic laboratories to maintain an acceptable high quality of their chromosome preparations. The better resolution (i.e. the higher number of bands) the smaller abnormalities may be detected. In some countries (e.g. USA and UK) committees have been established which control the quality of preparations from the cytogenetic laboratories. Such controls can only be sporadic and superficial and are not easy to quantitize. (See, however, the quality assessment procedure of Martin et al., this volume).

By using the multiple cell approach a great amount of data is available which can be used to assess the quality of cytogenetic preparations. The available measurements are shown in table 1. Some are obtained through the metaphase finding (low magnification) others during the automatic karyotyping (high magnification). It must be emphasized, however, that the measurements will only be useful if the system is adjusted and operated correctly.

**Table 1.** Measurements to be used for automatic assessment of the quality of cytogenetic preparations.

| Automatic measurements | Quality assessment |
|---|---|
| LOW MAGNIFICATION | |
| No. of located metaphases and scanning time | Mitotic index |
| Average ranking figure of located metaphases | Overall metaphase quality |
| No. of metaphases relocated to complete analysis | Quality of highest ranked metaphases |
| HIGH MAGNIFICATION | |
| No. of chromosomes classified into the composite karyotype and length and number of peaks of the density profiles | Band pattern quality |
| | Stage of chromosome contraction |
| | Metaphase spreading |

The total number of metaphases located on a certain area of the slide is a rough measurement of metaphase density. In the Magiscan system there is relationship between the cell density (dividing + undividing cells) on the slide and the time required for the automatic metaphase finding. By relating the number of located metaphases to the scanning time we may obtain an arbitrary measurement of the mitotic index. This measurement will probably only apply for rather clean preparations.

The average ranking figure of the located metaphases provides a measurement of their quality determined at low magnification.

The number and the average ranking figure of objects which are relocated in order to select the 10 metaphases required to complete the analysis provide measurements of the quality of the highest ranked metaphases.

By karyotyping measurements are available for assessment of metaphase quality at high magnification.

The length of the chromosomes classified into the composite karyotype provides information about the degree of chromosome contraction. The number of peaks in the density profile - which may have to be filtered to avoid noise (Lundsteen and Granum, 1978) - reveals information about the number of bands. Together these two measurements provide information about the quality of the chromosomes and their banding patterns.

The proportion of chromosomes which are classified into the composite karyotype mainly depends on how well the metaphase is segmented into isolated chromosomes by the automatic segmentation procedure. The segmentation procedure works best on well spread metaphases with relativly short chromosomes. By combination of the number of classified chromosomes and their degree of contraction it may be possible to obtain a measurement for the spreading of the metaphase.

By using these measurements or some of them regularly for all cases done at the laboratory, it is possible to monitor the quality fluctuations of slides produced at the the laboratory and it may even be possible to compare quality between laboratories which are using the same system.

## 3. Experiments and results

The success of the multiple cell approach to construct the composite karyotype is dependent of how many chromosomes are correctly classified. Enough pairs of homologues must be represented of each chromosome type in the composite karyotype in order to make it possible to detect structural abnormalities. The principle of the procedure was therefore tested in the following way:

From the daily workload of amniotic slides analysed on the system 25 random slides were selected. After automatic metaphase finding the operator counted 10 metaphases on each slide using the lightpen. Four of the metaphases were selected for automatic karyotyping without any use of the lightpen resulting in four hard copy karyotypes. As the multiple cell procedure was not yet implemented counting and karyotyping could not be done simultaneously. Therefore, the total time for each case as well as the time associated with the automatic karyotyping by the machine was recorded allowing calculation of the operator time actively spent at the machine. The chromosome pairs of the four hard copy karyotypes were cut out by scissos and arranged as a composite karyotype.

The average time for the full analyses of the 25 cases was 22 minutes of which an average of 9 minutes was spent by the computer making the karyotypes. Thus the average time spent by the operator working with the machine was 13 minutes ($\pm 2$ minutes). On the average additional 2 minutes were required to check each of the 25 composite karyotypes for structural abnormalities, so the total time per case was 15 minutes ($\pm 1$–2 minutes).

Table 2 shows the results of the automatic karyotyping of the 25 amniotic cases each represented by four karyotypes. On the average 1.6 of the chromosomes per metaphase which were classified into the composite karyotype were not properly segmented but looked anyway so much like "normal" chromosomes that they were not rejected by the classifier.

**Table 2.** Results of automatic karyotyping of 4 metaphases of each of 25 amniotic cell cultures without operator interaction.

| Case number | Classified but incorrectly segmented chromosomes No. | Correct segmentated chromosomes | | Correct classified chromosomes | | Correct classified pairs of homologues | |
|---|---|---|---|---|---|---|---|
| | | No. | % | No. | %* | No. | % |
| 1 | 2 | 36 | 78 | 32 | 89 | 10 | 43 |
| 2 | 0 | 40 | 87 | 37 | 93 | 15 | 65 |
| 3 | 1 | 37 | 80 | 33 | 89 | 13 | 57 |
| 4 | 2 | 40 | 87 | 39 | 98 | 17 | 74 |
| 5 | 2 | 35 | 76 | 30 | 86 | 11 | 48 |
| 6 | 1 | 36 | 78 | 33 | 92 | 12 | 52 |
| 7 | 2 | 38 | 83 | 35 | 92 | 14 | 61 |
| 8 | 1 | 40 | 87 | 38 | 95 | 16 | 70 |
| 9 | 1 | 37 | 80 | 32 | 86 | 11 | 48 |
| 10 | 1 | 33 | 72 | 27 | 82 | 10 | 43 |
| 11 | 2 | 38 | 83 | 34 | 89 | 12 | 52 |
| 12 | 1 | 40 | 87 | 38 | 95 | 16 | 70 |
| 13 | 2 | 39 | 85 | 38 | 97 | 16 | 70 |
| 14 | 3 | 36 | 78 | 32 | 89 | 12 | 52 |
| 15 | 3 | 30 | 65 | 22 | 73 | 6 | 26 |
| 16 | 3 | 30 | 65 | 24 | 80 | 5 | 22 |
| 17 | 1 | 37 | 80 | 36 | 97 | 14 | 61 |
| 18 | 1 | 37 | 80 | 33 | 89 | 12 | 52 |
| 19 | 1 | 39 | 85 | 37 | 95 | 15 | 65 |
| 20 | 2 | 38 | 83 | 36 | 95 | 13 | 57 |
| 21 | 2 | 34 | 74 | 29 | 85 | 10 | 43 |
| 22 | 2 | 37 | 80 | 34 | 92 | 13 | 57 |
| 23 | 2 | 34 | 74 | 30 | 88 | 10 | 43 |
| 24 | 1 | 37 | 80 | 33 | 89 | 11 | 48 |
| 25 | 1 | 37 | 80 | 35 | 95 | 13 | 57 |
| Average | 1.6 | 37 | 80 | 33 | 90 | 12 | 53 |

*Percentage of correctly segmented chromosomes.

On the average 37 chromosomes (80%) were correctly segmented per metaphase and 90% of these chromosomes were also correctly classified into the composite karyotype. On the average 12 pairs of homologues per metaphase were correctly classified which means that 53% of all 2.300 pairs of homologues were correctly ordered in the composite karyotype (range 22% - 74%).

# 4. Discussion

The purpose of this paper has been to describe an approach to automated cytogenetic analysis based on multiple cell analysis. Multiple cell approaches have been suggested previously by Granlund and coworkers (Granlund et al. 1976, Granlund 1978) and by Carothers et al. (1983) and discussed by Piper et al. (1980). A main difference, however between our procedure and the previous is that in the previous chromosome counting was automatic whereas our counting is done by the operator.

Correct automatic counting is difficult especially if low grade mosaicism has to be detected (Piper and Lundsteen, 1987). Consequently, many cytogeneticists may not be willing to rely on chromosome counting based on statistical analysis. Manual counting on the other hand is fast and reliable and provides an opportunity for the operator also to have a look at the chromosomes while counting.

The success of the proposed multiple cell approach to construct a composite karyotype clearly depends on how well the metaphases are segmentated and how well the chromosomes are classified. The results of the test showed that 80% of all 4600 chromosomes were correctly segmentated and that 53% of all pairs of homologues were correctly classified. Assuming that the probability for correct classification is identical for all 23 pairs of homologues one would expect that on the average 5 pairs of homologues for each of the 23 types would be present in the composite karyotype based on automatic classification of 10 metaphases. Consequently on the basis of our test using our current segmentation procedure and our current classification procedure on routine amniotic slides, we believe that the procedure may work in practice. Some hardware modifications are, however, necessary before the procedure can be implemented on the Magiscan, which at present is not able to execute more than one program at a time.

The multiple cell approach has some advantages but also a few possible disadvantages. Starting with the disadvantages it is clear that mosaicism of structural abnormalities may be overlooked. However, that kind of mosaicism may also be overlooked, if only 3 or 4 metaphases are karyotyped by the conventional method. Furthermore, the frequency of mosaicism of autosomal structural abnormalties is considered low, and regarding the sex chromosomes we routinely check these in all counted cells.

Another possible disadvantage is caused by the fact that the segmentation procedure works better on rather short chromosomes as compared to more elongated chromosomes. One could therefore expect the operator to prefer selection of metaphases with condensed chromosomes in order to ensure that enough chromosomes are segmentated and classified into the composite karyotype. In the past few years, however, much work has been devoted to the development of efficient segmentation procedures (see Graham, Ji, Wu and Vossepoel, respectively, this volume). It is therefore likely that segmentation may be improved so much that also more elongated chromosomes can be handled.

Among the advantages of the multiple cell approach the increase in the speed of the analysis is probably the most important. The test showed that the total time per case may probably be reduced by a factor of two resulting in an analysis time of approximately 15 min. per case. This implies that theoretically 32 cases might be analysed per 8 hour work day.

Another obvious advantage is that because no karyotyping interactions are required it is less tedious and tiring to operate the new system than the previous.

The procedure provides more flexibility because the operator is not forced to complete the analysis while sitting at the machine. The hard copy of the composite karyotype can be evaluated at any time, and by any member of the cytogenetic staff independent of the machine.

The composite karyotype is ideal for a fast and efficient search for structural abnormalities. Not only are the chromosomes arranged so that comparison between pairs of homologous chromosomes from different metaphases is facilitated. It is also expected that it will be the nicest chromosomes which will be classified into the composite karyotype.

A possible profit by using the multiple-cell procedure may be the automatic detection of minor structural abnormalities. We have thought of two different approaches, the statistical based on information from many cells and the single cell approach where information is obtained from what is expected to be the most informative pair of homologues selected from a number of cells. We have chosen the second approach, because small structural abnormalities may only be visible in elongated and well banded chromosomes, and because a possible small abnormality might well vanish, if it was to be detected on the basis of averaging of measurements from a number of chromosomes of which some might have poor or distorted banding patterns.

In the past evaluation and assessment of the quality of cytogenetic preparations have been based on subjective judgements and it is well-known among cytogeneticists that they do not always agree in their evaluation of the same material and that they do not always come to the same result, if they reevaluate the material after some time (Martin et al, this volume)

Automated chromosome analysis systems provide an opportunity to obtain quantitative, reproduceable measurements which are directly related to the quality of the cytogenetic material. A main difficulty is that the machine must always be well adjusted otherwise the measurements will change from time to time. We expect, however, that in the future systems will tend to be more robust and more reliable and therefore also useful for automatic quality assessment.

It is likely that automatic quality assessment will tend to increase the slide quality of the laboratory, because the continous monitoring of the quality of the preparations will draw the attention of the staff to any signs of degradation of the material.

In conclusion we expect that because of the increased speed and the possibility for automatic detection of structural abnormalities and for automatic quality assessment associated with the multiple-cell approach the metaphase finding + karyotyping systems may well become more cost effective than the karyotyping-only system in the future.

*Acknowledgements.* This paper was partially supported by the Commision of the European Communities through the Medical and Health Research Program, project no. II.1.1/13.

# References

1     Carothers AD, Rutovitz D, Granum E. An efficient Multiple-cell approach to automatic Aneuploidy screening. Anal Quant Cytol 5:194-200 (1983)
2     Granlund GH, Zack GW, Young IT, Eden M. A technique for multiple-cell chromosome karyotyping. J Histochem Cytochem 24:160-167 (1976)
3     Granlund GH. The structure of a system for multiple-cell chromosome karyotyping. In proceedings of the Fourth International Joint Conference on Pattern Recognition, Kyoto, Japan, pp.837-841 (1978)
4     Lundsteen C, Granum E. Description of chromosome banding patterns by band transition sequences. A new basis for automated chromosome analysis. Clin Genet 15:430-39 (1979)
5     Lundsteen C, Gerdes T, Maahr J. Automatic detection of structural chromosome abnormalities. In proceedings of the 5th European Chromosome Analysis Workshop, supported by the EEC, Heidelberg October 19-21, 1983.
6     Lundsteen C, Gerdes T, Maahr J. Automatic classification of chromosomes as part of a routine system for clinical analysis. Cytometry 7:1-7 (1986)
7     Lundsteen C, Gerdes T, Maahr J, Philip J. Clinical performance of a system for semi-automated chromosome analysis. Am J Hum Genet 41:493-502 (1987)
8     Lundsteen C, Martin AO. On the selection of systems for automated cytogenetic analysis. Am J Med Genet (In press, 1989).
9     Piper J, Granum E, Rutovitz D, Ruttledge H. Automation of chromosome analysis. Sign Proc 2:203-221 (1980)
10    Piper J, Lundsteen C. Chromosome analysis by machine. Trends in Genet 3:309-313 (1987)

# Towards a Knowledge-Based Chromosome Analysis System

Jim Piper, Richard Baldock, Simon Towers, Denis Rutovitz

## Summary

In this paper we review the problems of chromosome analysis systems built in the conventional pattern recognition and image processing paradigm, and suggest that what is required is a knowledge-based system approach to the control of such a system. In particular, evaluation of the performance of components of the system can lead to meaningful feedback strategies. We propose such a system in some detail in the context of SBS, a "blackboard system" shell. The system currently being constructed implements frame-based hypothesis handling and retains the experience invested in previous conventional (procedurally programmed) systems.

## 1. Introduction

Current automatic chromosome analysis systems are the result of about a quarter of a century of development of traditional pattern recognition and image processing (PRIP) techniques. They are both extremely complicated, and not very capable, requiring considerable operator interaction in order to perform satisfactorily. From both the intellectual and commercial viewpoints it would be desirable to improve these systems to be fully automatic; in which case, some important applications which cannot be economically addressed by either conventional automatic or interactive systems would become viable.

Chromosome analysis systems suffer from the typical problems of large, traditional PRIP programs; they are difficult to write, debug, or modify; and their behaviour in rarely exercised parts of the system may be unpredictable. So-called "Intelligent" Knowledge Based Systems (KBS) on the other hand are intended to incorporate knowledge in an accessible and explicit manner, with benefits of predictability, proper conformance to an underlying model, and easy incremental development. The extent that such good intentions will be realised in practise remains to be measured; meanwhile this new computing technology is accompanied by an extensive set of jargon. This paper considers the design of a KBS for chromosome analysis, paying particular attention to the problems of control of the analysis, and summarises the progress made thus far in its construction. We have two rather separate motives, firstly to substantially increase the capability of chromosome analysis systems beyond what has been achieved by conventional PRIP, and secondly to investigate the

**Automation of Cytogenetics**   Editors: C. Lundsteen J. Piper
© Springer-Verlag Berlin Heidelberg New York 1989

benefits and costs of KBS in a domain in which there has already been a lot of PRIP experience, and in particular where existing large data sets will allow objective measures of performance.

There have been earlier attempts to introduce more sophisticated control strategies for chromosome analysis. The Edinburgh MRC group have long proposed the use of *confidence measures* to evaluate processing decisions and possibly cause alternative processing paths (Hilditch, 1969). Granlund (1978) described a system which could alter its initial model of what feature measurements to expect, and thus in a multiple-cell system detect a cluster of abnormal chromosomes. Piper (1981) proposed a system for generating multiple hypotheses concerning centromere positions, with the "best" selected by a hypothesis evaluation method. Piper et al (1988) described a system for locating dicentric chromosomes in which hypotheses concerning centromere location were tested by taking further measurements from the original image; a form of "iconic model matching".

Recently, work on KBS chromosome analysis systems has started in several laboratories apart from our own; in particular Wu *et al* (1987) have published a critique of traditional chromosome analyis and a conceptual outline of a production rule based system which might provide a solution; this concept is being actively developed and extended (Wu, personal communication), while the Manchester group have experimented with alternative model-based strategies for chromosome segmentation (Thornham *et al*, 1988, Graham and Taylor, 1988, Cooper *et al*, 1988). In this paper we describe in some detail the knowledge representation, knowledge sources, and computation mechanisms which we are beginning to use to tackle the chromosome analysis problem. In particular, we describe in detail the evidence and feedback management which Wu *et al* (1987) believed would be necessary but for which they made no detailed proposals. Chromosome analysis is used for a wide variety of clinical, screening and biological measurement purposes. Our system will initially concentrate on the clinically important task of constitutional karyotyping, which is the description of cells in terms of the normal chromosomes contained therein and in particular the detection and characterisation of any abnormality.

Despite taking a new approach, we have decided to exploit where appropriate the very substantial investment of programming effort in our earlier systems for two reasons. Firstly, conventional procedural programming is always likely to be required for the image-based processing in order to obtain reasonable computational efficiency. Secondly, notwithstanding the overall limitations of their capability, many existing systems actually work rather well much of the time, and the behaviour of their individual modules is well understood; what is missing is supervisory evaluation and control so that the system can recognise errors and investigate alternative solution pathways.

The first important task when building a KBS is to make explicit the models to which the system is trying to match its images. In some areas of vision research it is possible to use a *synthesis model* which describes how an image is produced from, for example, an illuminated three-dimensional subset of the world and a camera, to guide the construction of an *analytic model* which describes the generation of knowledge about the three-dimensional world from particular images. For example, one can think of projective geometry applied to a "blocks world". However, we do not know how to construct a useful synthesis model of chromosome image formation and we expect that our knowledge of how chromosome images are formed will only be of use in providing constraints for an analytic model. The causes of chromosome image variability and the underlying invariant properties of chromosomes and cells are discussed in section 2. There has in fact been a long history of implicitly model-based or "principled" PRIP applied to chromosome analysis, with varying degrees of success, and which may well contribute to a KBS implementation. Some of these methods are discussed briefly in section 3.

The other important task is to decide and configure for use a suitable system architecture. Our proposed system will comprise a number of separately implemented knowledge sources or "experts" in particular subdisciplines communicating via a "blackboard". The philosophy of such a system is introduced and the overall structure is described in some detail in section 4. Section 5 summarises the status and future plans for the project.

## 2. Causes of Chromosome Image Variability

We shall for simplicity restrict the discussion to human chromosomes. Each normal cell contains 46 chromosomes drawn from 24 classes, with 22 autosomal pairs and two sex chromosomes, XX in females and XY in males. At the metaphase of cell division the chromosomes contract to form objects typically between $1\mu m$ and $2\mu m$ wide and between $2\mu m$ and $15\mu m$ long. The chromosomes do not all look alike; within a cell there is a range of sizes, morphology and banding pattern (figure 1). The banding pattern does in fact depend significantly on the particular preparation technique used, but for brevity we will confine discussion to a single method.

Normally the pair of chromosomes in a class are homologues of each other and of the corresponding chromosomes in other cells, and usually differ biologically only in certain small and well characterised features. However, *visual* appearance is very varied. We are dealing with rather elastic or "floppy" three dimensional objects which are stabilised and flattened by being summarily deposited on and stuck to a glass microscope slide. Nevertheless, there *are* more or less invariant properties of the visible chromosome features, notably the relative size within the cell, the relative position within the chromosome body of the centromere (a well-marked constriction in the typical axially bi-symmetric chromosome shape), and the banding pattern associated with each class. In more detail, the major variations comprise:

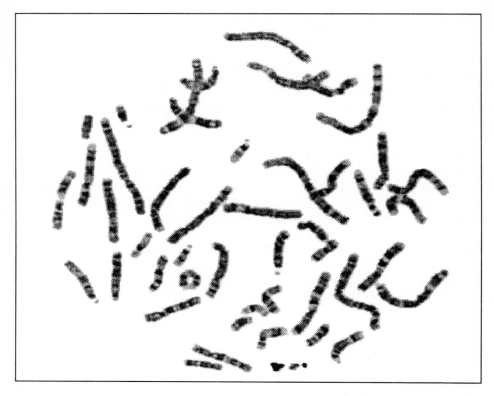

**Fig. 1.** A digitised and thresholded G-banded metaphase cell, showing bent, touching and overlapping chromosomes, and at the bottom some non-chromosomal stain-debris.

1. Location. Position and orientation are in general random, excepting in so far as in selected "good" preparations most chromosomes are completely separated from the others; a few form clusters of touching chromosomes, and a very few overlap (occlude) other chromosomes (figure 1).
2. Chromosome deformation. In the course of deposition on a slide, chromosomes may bend, twist, shrink or stretch differentially.
3. Biological variability. The visual appearance of a particular specimen depends in a way which is far from being well-defined on the individual, the tissue, and the circumstances of the sample preparation.
4. Contraction effects. Not all cells are precisely at the same stage of cell division, and thus not all chromosomes are equivalently contracted. This has four main consequences. Firstly, absolute chromosome size is of little classificatory significance, though relative size within each cell is a very strong feature. Secondly, the number of bands expressed on a chromosome depends on the state of contraction; and is thus a relative rather than an absolute feature. Thirdly, chromosome shape and particularly the

appearance of the centromere is very much dependent on contraction; in less contracted cells the centromere may not be visible; its position being only deducible from the band pattern. Fourthly, longer chromosomes are more often and more severely bent or overlapped than shorter ones.

5. Abnormality. Chromosome analysis is ultimately concerned with detecting abnormality, which usually presents as *aneuploidy* (abnormal chromosome number), or as structural rearrangement of some chromosomes, either randomly or consistently between cells.

There are some constraints on variability which are used in human visual analysis. The most important are the within-cell constraints concerning relative size (above), the relative number and contrast of the bands, and the obvious visual similarity between the homologous chromosome pairs. The within-sample constraints relate to preparation quality, and the fact that for the majority of cases, constitutional chromosome abnormality is present identically in every cell examined, though there are clinically significant varieties of chromosome damage in which this is not the case.

# 3. Analytic Models for Chromosome Analysis

Owing to the variability described above, it is not really possible to construct and use a chromosome generation model as the basis of chromosome analysis. However, one may proceed from a description of a cell in terms of the chromosomes present, their classes, the preparation technique and the stage of metaphase at which the cell has been fixed to an abstract description in terms of chromosome size, centromere position and band pattern which is quite precise; it may be represented for example by the ISCN templates (figure 2) (ISCN, 1985). This transformation could be computed, and presumably the reverse process is also soluble. However, going from this abstract representation to the actual observed image involves geometrical transformations which are very difficult to characterise; the situation is entirely unlike the common vision problem where the transformations (projective geometry, Lambertian surface illumination, occlusion) are deterministic from the 3-D representation to the 2-D image. The main aim of chromosome analysis being the detection of abnormalities, either the analytic system must be capable of detecting any departure from the "normal" model, or, preferably have access also to a model of the various abnormalities of structure and number which may be encountered, with a concurrent ability to characterise abnormalities detected in a cell, and in some cases, to carry forward hypotheses of abnormality from one cell to another. Nowadays these seem challenging, but not unobtainable objectives. Hitherto however, chromosome analysis systems have always been based on attempts to build a "bottom-up" description of the image. Some approaches that have been used with at least partial success are briefly described in the following paragraphs.

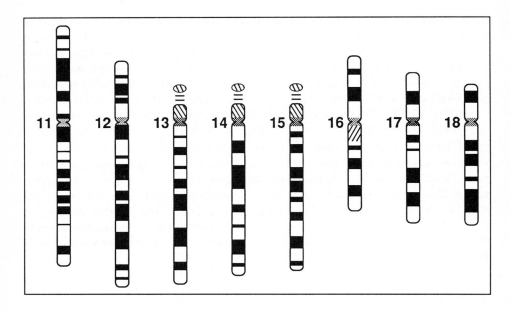

**Fig. 2.** Examples of the ISCN 550-band templates.

Initial segmentation has almost invariably been implemented by finding connected above-threshold image components, local or global thresholds being chosen by an analysis of pixel density value histograms (Mendelsohn *et al*, 1969, Rutovitz, 1977). The resulting objects are usually single chromosomes, but clusters of touching and overlapping chromosomes are common, as is non-chromosomal material such as an undividing nucleus or stain debris (figure 1). Various attempts have been made to classify the segmented objects as chromosome or one of the possible non-chromosome types by shape analysis, and then re-segment clusters using boundary analysis and valley-following techniques (Ledley *et al*, 1968, Gallus and Neurath, 1970, Vanderheydt *et al*, 1981, Vossepoel, 1987, Ji, 1988, Wu *et al*, 1988).

Features that are relatively *invariant* to the random nature of chromosome position, contraction and bending can be measured using a non-Euclidean coordinate system determined by the chromosome's symmetry axis, which can be found with more or less success by a number of methods (Piper *et al*, 1980).

As was indicated in section 2, most chromosome shape variation is not significant for recognition. However, the location of the centromere and telomeres (tips of the arms) is important and a variety of approaches have been used, using either boundary curvature analysis (Rutovitz, 1969, Ledley *et al*, 1972, Gallus and Neurath, 1970) or analysis of a single dimensional projection of some property of the chromosome onto the symmetry axis, e.g. of width or integrated density (Mendelsohn *et al*, 1969), or a combination of both (Piper, 1981, Piper *et al*, 1988).

Size and shape can discriminate the 46 human chromosomes into 10 groups. To refine the classification into the full 24 classes and to improve the resolution of abnormality, use of staining regimes which result in a typical banding pattern for each chromosome class is now universal. There are two paradigms for automatic analysis of the patterns, either by building a "local" description based on each band (Granlund, 1973, Lundsteen and Granum, 1979, Oosterlinck *et al*, 1977, Thomason and Granum, 1986), or by extracting a variety of "global" descriptors, for example Fourier components of a one dimensional profile of density (Caspersson *et al*, 1971, Granum, 1982).

Typically chromosomes are classified in isolation, and then the overall classification of the cell is modified to satisfy a few model-determined constraints. Most isolated or unconstrained classification relies on one of a variety of statistical models of the within-class and between-class feature distributions (e.g. Ledley *et al*, 1980, Granum, 1982, van Vliet *et al*, 1988). Constraints on what classification results are possible or likely are potentially powerful, but in practise only four have been applied:

1. The likelihood of a chromosome belonging to a class must be reasonably large, otherwise the object should be treated as a potentially wrongly segmented or abnormal chromosome.
2. Nearly all classes should contain exactly two chromosomes.
3. Chromosomes from the same class should be generally rather similar (Granlund, 1978, Jaschul, 1985, Zimmerman *et al*, 1986).
4. Constitutional abnormalities of the karyotype are usually present in every cell examined, and so "composite karyotyping" which integrates results from a number of cells can lead to an overall correct diagnosis (Hilditch, 1969, Carothers *et al*, 1983, Granlund, 1978).

Various usually rather *ad hoc* methods have been used to try to accomodate these constraints - mostly amounting more or less formally to a relaxation constraint satisfaction process (Hilditch, 1969, Rutovitz, 1977, Granlund 1978, Piper, 1986, Vanderheydt *et al*, 1979).

# 4. KBS Control of Chromosome Analysis

The individual components of current chromosome analysis systems surveyed above function rather well when presented with appropriate data, but fail when the input data does not accord with their implicit model, usually because some of the variability described in section 2 has not been adequately taken into account. Thus for example, segmentation by thresholding works faultlessly on those rare, ideal cells which have no touching or overlapping clusters of chromosomes. Statistical classifiers work well when presented with correctly measured feature values from chromosomes "similar" to the training set; feature measurement generally works well when a chromosome is well segmented and its correct symmetry axis has been found.

In order to deal with these problems a "control" capability is required that incorporates evaluation of the suitability of input data; propagation of "confidence" measures in the results of particular processes to subsequent analysis stages; ability to reason with the corresponding uncertain evidence; verification of hypotheses by appropriate "iconic" or image-based rules; appropriate feedback when input or result confidence is low; and other order-of-processing desiderata. Also desirable is the ability to incorporate knowledge incrementally; and to incorporate in particular that control knowledge which can best be expressed symbolically rather than algorithmically or numerically, for example the rules expressed informally in figure 3. We expect that the major use of rules will be to control the evaluation and feedback within the system, rather than to replace algorithmic knowledge. However, the distinction will obviously be blurred in some cases.

---

**if**      *there appear to be 45 chromosomes in the cell, with an additional*
            *C group chromosome, but a missing D and a missing G*

**then**    *suspect a missed touching cluster of a D and a G*

**unless**  *the pattern is repeated in all cells in the sample, in which case*
            *suspect a Robertsonian translocation*

---

**Fig. 3.** An informally stated rule which uses model knowledge to assist with a segmentation decision.

With these considerations in mind we can identify two main forms of control knowledge which should be represented within the system:

1. Consider a "module" such as segmentation. Clearly there is knowledge concerning what such a module can do, what inputs it requires and what prerequisities must be satisfied, which is used to determine when to invoke the module.

2. The rule in figure 3 is part of the knowledge about how to test the hypothesis that the present cell classification is correct. Clearly, production rules are suitable for knowledge of this kind. Knowledge of this type would seem to more naturally fit "within" a module, rather than defining the use of the module as in the previous case.

Amongst the possible approaches to KBS control, the so called "blackboard systems" have been favoured by workers in diverse fields ranging from speech recognition (Erman *et al*, 1980) to guidance of underwater vehicles (Russell and Lane, 1986). A blackboard architecture consists of at least three elements: a shared data region (called the blackboard), a set of "experts", and a scheduler (Ensor and Gabbe 1986). The experts contain "rules" and "procedures" which express the domain expertise of the system. The scheduler controls execution of the experts according to information on the blackboard. The inherent modularity of such an architecture is particularly useful for systems which are attempting to solve problems with complex information interdependencies and diverse knowledge types. Furthermore, the modularity of the experts and formalisation of the communication means that it is easy to install or remove experts. We have chosen to use this type of system, and the rest of this section outlines the major factors for consideration when implementing a control model on a "blackboard" system.

Other software architectures for KBS are possible. For example, production rule systems are widely used. However, the use of a blackboard does not preclude the use of such architectures within individual experts, but does permit greater flexibility, particularly with regard to incorporating existing useful procedures. Thus the blackboard is simply a way of organising the various modules and their communication and does not in itself provide a solution to the image processing task.

## 4.1. SBS, a blackboard system shell

SBS (Baldock, Ireland and Towers, 1987) is a blackboard system shell written in POP-11 (Barrett *et al*, 1985) in the POPLOG environment (System Designers, 1986), which also allows easy access to C and any other compilable code. Figure 4 is a schematic diagram of SBS where the arrows indicate lines of communication.

Operation of the system can be summarised as follows. There is a goal list which acts as an attention focussing mechanism and gives the system designer some explicit control over the logical order of execution of the experts. An expert may or may not be able to assist the achievement of any particular goal, if it can, it is said to be able to "satisfy" the goal. The system attempts to satisfy the goal at the head of the goal list. Initially the list is empty and the user is asked to enter a goal. After this experts add or remove goals (see below), until either the list becomes empty (indicating completion of the task), or the scheduler detects that none of the experts can satisfy any of the remaining goals; the task has failed.

Each expert has two parts. The first is an enumeration of the goals which the expert can satisfy and their corresponding prerequisite conditions, and actions in the event of failure to meet the conditions. The second part contains the set of actions to be executed by the expert in the event that the prerequisite conditions are satisfied. The actions comprise a set of rules and/or procedures. Execution of an expert is correspondingly a two stage process:

1. The first part of the prerequisite, corresponding to the goal being considered, is tested to determine whether the conditions for satisfactory application of the expert's actions are satisfied. If the prerequisite conditions

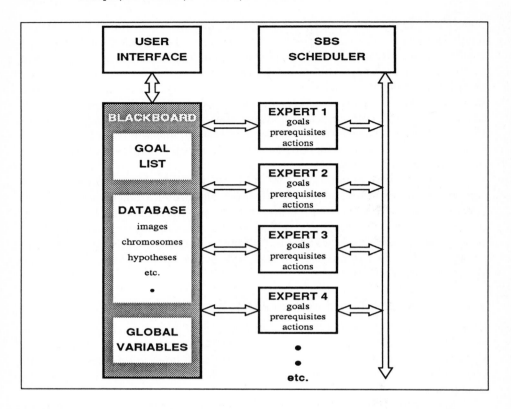

**Fig. 4.** Schematic diagram of SBS and the relations between experts and blackboard.

are not satisfied, the prerequisite actions are executed causing sub-goals to be added to the goal list. Execution of the present expert is terminated and the scheduler attempts to select another expert to satisfy the new sub-goals. If the prerequisite conditions are satisfied the system goes on to execute the actions of the expert.

2. The actions of the expert involve running a production rule mechanism on a user-defined rule base. The rules are ordered according to a priority parameter which may be changed dynamically. Processing control remains within the expert until no rules can fire or until an explicit "exit" call is executed. On exit, the goal that caused invocation of the expert is regarded as "satisfied" and so removed from the goal list. It should be noted that if the actions of another expert are required from within a rule base, then this can only be achieved by putting the appropriate goal on the goal list, and quitting the current expert.

Those familiar with expert systems terminology will see from the above that the experts' prerequisites provide the backward-chaining and planning

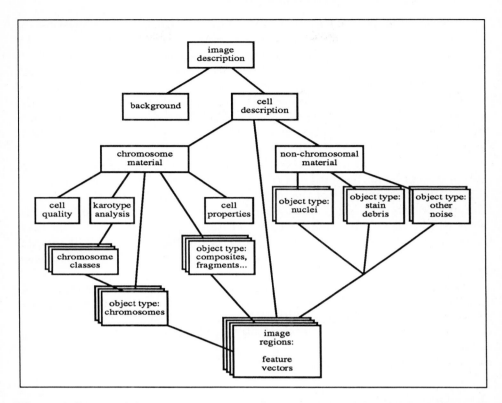

**Fig. 5.** A diagram of the major frame types used to model the analysis of a metaphase cell and their most important connections. This network will ultimately be extended upwards to handle multiple cell knowledge, and downwards to provide a frame basis for image region feature measurement.

aspects of the control mechanism, while the experts' actions provide the forward chaining and procedural parts; we observe that since the rule base associated with an expert need comprise only one rule whose "action" part could simply execute procedures, normal procedural control is incorporated.

## 4.2. Knowledge representations

The word *frame* has been applied to a variety of "slot-and-filler" declarative representation structures. In this work we do not consider a calculus for reasoning in the context of frames (Minsky, 1975), but simply use them as a convenient data structure. In particular we use an implementation of frames (Towers, 1987) in the SBS system for abstractions and interpretations of the data and to record the progress of the analysis (hypothesised chromosomes, clusters, etc., chromosome classes, hypothesised cell descriptions, processing histories).

The network of frames used to describe an image of a metaphase cell is shown in figure 5. In this figure, the lower frames are "components" of higher ones (Baldock *et al*, 1987). During the analysis of an image, instances of these frames are constructed and placed in the SBS blackboard, and subsequently manipulated by the experts or "knowledge sources". Figure 5 also indicates where multiple instances of a frame will occur in the system; for simplicity only a single instance of a *cell description* is shown. In fact, the entire system which we currently envisage, including multi-cell karyotyping capability, is somewhat more extensive than that shown in figure 5.

Hypothesised cell descriptions play a crucial role in the system and each is represented as an instantiation of a frame network. These describe a partition of the "foreground" of the input image into hypothesised chromosomes, chromosome clusters, and other non-chromosomal material, and provide the basis within which reasoning about segmentation and classification can take place. In all likelihood, many such hypothetical cell descriptions will have to be considered by the system; however they will share most of their components most of the time.

Image and graphic data structures (segmented objects, symmetry axes, profiles) are stored as appropriate data structures in the MRC "Woolz" C-language image analysis system (Piper and Rutovitz, 1985), and image processing procedures are written in C. Symbolic rules are implemented as production rules in the SBS system, and some associated procedures in the experts are written in POP-11.

## 4.3. Knowledge sources and experts

Our earlier PRIP chromosome analysis system contained a number of "modules" each responsible for a particular *domain* task, for example segmentation, in which knowledge is represented exclusively in procedural form. It seems natural that in SBS each of these modules should be associated with an individual *expert* or *knowledge source*. While the procedure bodies can be retained more or less as they are, considerable programming effort was required to implement the KBS control aspects of the system. The major knowledge sources are summarised in figure 6; several are primarily concerned with control activity, e.g. given a primary goal, identify of the necessary sub-goals. More details of the prerequisites and actions of the knowledge sources are described in Piper *et al* (1987).

## 4.4. Confidence and evidential reasoning

Execution starts by building a cell description by "bottom-up" image processing and initial classification. The system then enters a "hypothesize-and-test" control mode. To compare different hypotheses we must be able to generate a *degree of belief* or confidence for each. Most of the contributing evidence

| KS1: | background-foreground expert: this computes a threshold, and places new frames for connected above threshold *image regions* in an *image description* frame which is put on the blackboard. |
|---|---|
| KS2: | cell selection expert: measures the acceptability of a set of image regions for further analysis. |
| KS3: | image enhancement expert: not yet implemented. |
| KS4: | image region feature measurement expert: This expert embodies a large amount of algorithmic structural image analysis, for example axis and centromere location. |
| KS5: | "object typing" expert: computes vector of object type probabilities by a combination of statistical and rule-based structural analysis. |
| KS6: | feature normaliser expert: normalise feature sets in the context of a complete cell |
| KS7: | statistical chromosome classifier expert: generate a "class likelihood vector" that relates an *image region* to each *chromosome class*. |
| KS8: | karyotype construction expert: assign each *image region* to a normal chromosome class, or to an "abnormal" class, using a cell-wide global best fit subject to various constraints. See §4.4, §4.5. |
| KS9: | karyotype evaluation and hypothesis formulation expert: generate alternative hypotheses concerning (i) object typing, (ii) chromosome class, (iii) constitutional karyotype of sample, etc., by considering (a) knowledge from other cells, (b) object type and feature confidences, (c) overall likelihood of classification, (d) ad hoc rules. |
| KS10: | split/merge regions expert. See §4.6. |
| KS11: | hypothesis generation and management expert. See §4.6. |

**Fig. 6.** The division of the chromosome analysis system into knowledge sources or "experts"

which supports or conflicts with a hypothesis is uncertain in nature, for example classifier likelihoods, segmentation ambiguities or feature measurement confidence. We need to be able to combine such evidence, possibly conflicting, from diverse sources in a theoretically sound manner. Cheeseman (1985) argues convincingly that it is sound practice to combine degrees of belief by use of the formulae of probability (addition, multiplication, Bayes' theorem, etc.). The difficulty is that the available information is generally not sufficient to determine conditional probabilities, i.e. the problem is under-constrained. A common way around this difficulty is to make additional assumptions to provide additional constraints, for example to assume "conditional independence"

(Kim and Pearl 1983) which allows the factorisation of the conditional probability matrix into terms of much smaller dimensionality. If it is not possible to
determine which terms of the problem are conditionally independent then it is
sufficient to assume maximum entropy which will determine the "most honest"
distribution for a given set of data (Cheeseman 1983).

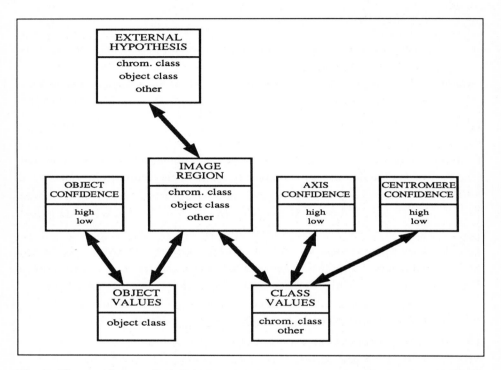

**Fig. 7.** The causal network of a single image region. Each node can exist in a number of
states, summarised in the lower box of each node, and records causal and diagnostic evidence
contributing to the current degree of belief or confidence for it to be in any one of the states.
The arrows indicate that each link is bidirectional

In the system reported here, evidential reasoning about classifications of
image regions as chromosomes of particular classes or other object types is being
implemented in expert KS8 using the probabilistic model of Kim and Pearl
(1983). Currently, each region has a network similar to that shown in figure 7,
which is a sub-graph of the inference net we believe is ultimately required.
Because of the inherent modularity of such networks it is straightforward to add
more nodes to include evidence from other sources. Essentially, the confidence
of a particular interpretation at a node in the network is functionally dependent
on causative and evidential inputs from neighbouring nodes in the network.

The values at nodes may also depend on the prior or external information which could arise from measurements, externally imposed constraints, or from a higher level hypothesis. Change in any belief value at any node (e.g. on account of a measurement having been made) results in propagation of calculations through the network with new beliefs and evidence values computed at each node. The main advantage of this approach is that it is locally computed; i.e. confidence at a node knows nothing about other nodes in the network except for the evidential inputs generated by immediate neighbours. Furthermore, provided the network is acyclic then the system is well behaved and a change at a node propagates through the network in a single pass.

As an example, consider a frame representing an image region which might be a cluster or composite of two or more chromosomes not so far segmented. Internal evidence that the region is a cluster will come, say, from a statistical object type classifier, which produces statistically valid probabilities *on the basis of measured object features alone.* Subsequently, evidence of the region being a composite may also be provided by a higher level interpretation failing to classify it as a single normal chromosome; and when the region is passed to a resegmentation expert, more evidence for or against the cluster hypothesis will be generated by the success or failure of the resegmentation, and if successful, by the subsequent evaluation of the regions which result from the resegmentation. So far as the individual image region in question is concerned, all of these aspects of confidence in its interpretation, as well as the confidence in the feature measurements (some of which are computed by complex, error prone and evaluable software), can be handled by the image region's causal network (figure 7).

### 4.5. Chromosome classification constraint management

Classification constraints can be expressed as rules for modifying the state values in the "external hypothesis" nodes of the causal networks (figure 7). We are implementing an iterative relaxation constraint-management scheme in which expert KS8 uses rules to modify the external hypothesis inputs, and then employs the causal networks, to construct a classification of the image regions which in some sense globally maximises the satisfaction of constraints between the chromosomes.

Typical constraints are e.g. (i) chromosomes whose class-mean size is "big" are usually larger than those whose class mean-size is "small" (but not always), (ii) a chromosome X has a larger centromeric index than a chromosome 7 (but not if the centromere finding algorithm made a mistake), (iii) two chromosomes from the same class should have nearly identical feature values (but not necessarily if one has suffered some unexpected process e.g. differential stretching). Other constraints can be expressed in terms of the expected number of chromosomes in each class, the class likelihood values, evidence from other cells, prior knowledge, etc..

- **if** a *likely class is 4* **and** *there is a prominent dark band on arm q adjacent to the centromere* **then** *increase likelihood* **else** *decrease likelihood*
- **if** a *likely class is 6* **and** *arm p is on average paler than arm q* **then** *increase likelihood* **else** *decrease likelihood*
- **if** a *likely class is 7* **and** *there is a dark band at the tip of arm p* **then** *increase likelihood* **else** *decrease likelihood*
- **if** a *likely class is X* **and** *the most prominent dark band lies 30% along arm q* **and** *there is a prominent dark band on arm p equi-distant from the centromere* **then** *increase likelihood* **else** *decrease likelihood*
- **if** a *likely class is 13* **and** *the top half of the chromosome is paler than the bottom half* **then** *increase likelihood* **else** *decrease likelihood*
- **if** a *likely class is 15* **and** *the top half of the chromosome is darker than the bottom half* **then** *increase likelihood* **else** *decrease likelihood*

**Fig. 8.** Some rules for using "iconic" model matching to confirm hypothetical classifications of chromosomes

A particularly important source of additional information comes from "iconic" model matching; i.e. looking directly at the image for features which confirm a current hypothesis about an image region. The original, hypothetical classification of a chromosome generated by the statistical classifier is made by computing a formula which incorporates a number of normalised features values, namely size, centromeric index, and global band pattern features (Piper and Granum, 1988). Misclassifications can arise for a variety of reasons, in particular, incorrectly located centromere, or miscomputed normaliser constants for whatever reason. The aim here is therefore to make measurements from the image which are somewhat independent of the feature set used for the initial classification, and an obvious source of such information is to look at structural "landmarks" on the chromosomes, which the global band pattern features do not necessarily represent satisfactorily, particularly in longer chromosome preparations. A few of the many possible rules of this type are shown in figure 8. Several points are noteworthy. Firstly, the "condition" part of these rules deliberately restricts their application to chromosomes currently believed likely to belong to a particular class; if rules of this nature were applied in an unconstrained fashion then a large number of inappropriate chromosomes might match a particular rule. For example, the rule concerning class 4 is equally applicable to class 10 chromosomes. Secondly, each rule requires appropriate image-level procedures to make the match; e.g. prominent band extraction by Gaussian decomposition (Granlund, 1973). Thirdly, the rules specified are specific to metaphase chromosomes, and different rules may be required for longer chromosomes. Clearly the rule base which is currently being built will in the end be rather large, and evaluating overall consistency, and the utility of each rule, will be a major undertaking.

### 4.6. Generation and evaluation of alternative hypotheses

The overall confidence attached to a cell description arises from the degrees of belief in the chromosome and other object type classifications obtained from the image regions' networks. If the overall confidence is low, the cell evaluation expert KS9 investigates likely reasons for constraints not to have been satisfied, e.g. apparently abnormal chromosome number, or apparent presence of an abnormal chromosome. In such an eventuality it uses a rule-base to generate alternative hypotheses. These will correspond to one of the following actions, (i) accept that the cell is indeed abnormal (which results simply in a duplication of the top-level cell description frame with an altered slot value on normality resulting in a higher overall confidence), (ii) suggest re-classification of one or more chromosomes, (iii) suggest re-classification of the object type of an image-region, in other words reconsider the segmentation.

We will consider the third possibility as being the most complex in subsequent behaviour. In such a case, KS9 enters a slot value for the "probability of being composite from karyotype evaluation" in a *copy* of the appropriate image region frame, and sets a goal "generate new hypothesis around image region copy". KS11 makes a new cell-description frame in which every sub-frame *except the image region specified in the goal* is a copy of those in the original (or, preferably, where ambiguity cannot occur, a link to the original version), and the image region specified on the goal list is incorporated with its new evidence values.

The new evidence when taken with the existing evidence from object-type features *may* be sufficient to cause re-interpretation of the image region as a composite. In this case, segmentation is attempted by KS10. Successful segmentation reinforces the confidence of the object type being composite, while failure obviously decreases the confidence. Processing of the new cell description proceeds with feature measurement of image regions resulting from splitting, renormalisation if deemed necessary, statistical classification, and karyotyping. At the end, the new cell description frame ends up with a possible karyotype together with an overall confidence.

There are now two cell descriptions. Each can be evaluated and each may spawn yet further hypotheses. KS11 is able to recognise duplication of hypotheses resulting from alternative paths through intermediate hypotheses. Hypotheses which "made things worse" are generally not used as the basis for generating further hypotheses, both because they do not provide a plausible basis for such an operation, and to prevent combinatorial explosion. Low-confidence interpretations must however be retained since their confidence may later be increased by additional top-down knowledge. An obvious example would be when the sample is abnormal and this fact is recognised only after failing to reach a high-confidence normal interpretation in each of several different cells.

If the confidence that the object that was the subject of the hypothesis never reached the threshold at which it was accepted as a composite (possibly because the object-type feature evidence was overwhelming, or possibly because KS10 failed to split it) then the new cell description hypothesis will inevitably obtain a low overall confidence – an "unsplitable composite" is not good news.

## 5. Summary and Future Plans

In this paper we have described in some detail a plan for a knowledge based approach to the control of the procedures of chromosome analysis. In particular, we described the strategy for knowledge represension, some detail of the knowledge sources, and the computational engine. Some emphasis was placed on the problems of confidence and evidence management, and a feedback strategy based on the generation and evaluation of alternative hypotheses was proposed. Implemention of this system is well under way; goal-directed bottom-up processing that mimics existing PRIP systems is already functioning; the causal network and constraint management is being built; alternative hypotheses can be set up manually and will then run. Much remains to be done, in particular the building of a rule-base and associated iconic feature measurement to provide class-specific model matching in KS8, and the KS9 rule base for the generation of alternative hypotheses.

We hope soon to be able to start testing the efficacy of the KBS system, using the various data-bases of classified chromosomes which have been collected over the years by various groups; this will provide a unique opportunity to compare a KBS objectively with conventional PRIP systems.

Later, we intend to address possibilities of parallel processing and the extension of the knowledge based system to include the low level processing, particularly chromosome axis and centromere location, and multiple cell analysis. So far as parallelism is concerned, there is ample scope for parallel execution of the system outlined above. There is natural parallelism in the processing of image regions, of alternative cell descriptions, and especially, the execution of different experts. Integrating knowledge from a number of cells is clearly important in constitutional chromosome analysis. This will be included as an additional knowledge source at a later stage and will require modifications to confidence computation procedures in KS8, and to KS9. The relative ease or otherwise of this modification will be an interesting test of the flexbility of the KBS approach and the SBS implementation.

*Acknowledgements.* We are grateful for numerous helpful comments from and discussions with various participants in the EC Concerted Action on Automation of Cytogenetics, particularly Jim Graham and Qiang Wu. George Spowart wrote the PostScript program which drew the ISCN chromosome templates in figure 2. Muriel Lee and Eric Thomson provided expert assistance with the elucidation of the rules in figure 8.

# References

1. Baldock RA, Ireland J, Towers SJ. SBS User Guide. MRC Human Genetics Unit, internal report (1987)
2. Barrett R, Ramsay A, Sloman A. POP-11 - A Practical Language for Artificial Intelligence. (Ellis Horwood, Chichester, 1985)
3. Carothers AD, Rutovitz D, Granum E. An efficient multiple-cell approach to automatic aneuploidy screening. Anal. Quant. Cytol. 5:194-200 (1983)
4. Caspersson T, Lomakka G, Moller A. Computerised chromosome identification by aid of the quinacrine mustard fluorescence technique, Hereditas 67:103-109 (1971)
5. Cheeseman PC. A Method of Computing Generalised Bayesian Probability Values for Expert Systems. Proc. 8th Nat. Conf. A.I., Washington. pp.70-73 (1983)
6. Cheeseman PC. In Defense of Probability, Proc. 9th I.J.C.A.I., UCLA. pp.1002-1009 (1985)
7. Cooper DH, Bryson N, Taylor CJ. An Object Location Strategy using Shape and Grey-level Models. Proc. 4th Alvey Vision Conf., Manchester. pp.65-71 (1988)
8. Ensor JR, Gabbe JD. Transactional Blackboards. Art. Intell. in Eng. 1:80-84 (1986)
9. Erman LD, Haye-Roth F, Lesser VR, Reddy R. The Hearsay-II Speech-Understanding System: Integrating Knowledge to Resolve Uncertainty. ACM Computing Surveys. 12:213-253 (1980)
10. Gallus G, Neurath PW. Improved computer chromosome analysis incorporating pre-processing and boundary analysis. Phys. Med. Biol. 15:435-445 (1970)
11. Graham J, Taylor CJ. Boundary Cue Operators for Model Based Image Processing. Proc. 4th Alvey Vision Conf., Manchester. pp.59-64 (1988)
12. Granlund GH. The use of distribution functions to describe integrated density profiles of human chromosomes. J. Theor. Biol. 40:573-589 (1973)
13. Granlund GH. The structure of a system for multiple-cell chromosome karyotyping. Proc. 4th I.J.C.P.R., Kyoto, Japan. pp.837-841 (1978)
14. Granum E. Application of statistical and syntactical methods of analysis and classification to chromosome data. In: Kittler J et al (eds). NATO ASI series no. C.81: Pattern Recognition Theory and Applications. (D. Reidel, Dordrecht, Netherlands, 1982) pp.373-398
15. Hilditch J. The principles of a software system for karyotype analysis. In: Jacobs PA, Price WH, Law P (eds). Human Population Cytogenetics. (Edinburgh University Press, 1969) pp.298-325
16. ISCN. An international system for human cytogenetic nomenclature (1985). (S. Karger, Basel, 1985)
17. Jaschul H. The IBAS as a chromosome analysis machine. Preliminary Report on EEC Work Group Meeting on Automated Chromosome Analysis, Leiden, October 1985, pp.116-124
18. Ji L. Intelligent splitting in the chromosome domain. Pattern Recognition, (in press)
19. Kim JH, Pearl J. A computational model for causal and diagnostic reasoning in inference systems. Proc. 3rd I.J.C.A.I, Karlsruhe pp.190-193 (1983)
20. Ledley RS, Ruddle FH, Wilson JB, Belson M, Albarran J. The case of the touching and overlapping chromosomes. In: Cheng GC, Ledley RS, Pollock DA, Rosenfeld A (eds). Pictorial Pattern Recognition. (Thompson, New York, 1968) pp.87-97
21. Ledley RS, Lubs HA, Ruddle FH. Introduction to Chromosome Analysis. Comp. Biol. Med. 2:107-128 (1972)
22. Ledley RS, Ing PS, Lubs HA. Human Chromosome Classification using Discriminant Analysis and Bayesian Probability. Comp. Biol. Med. 10:209-218 (1980)
23. Lundsteen C, Granum E. Description of chromosome banding patterns by band transition sequences. A new basis for automated chromosome analysis. Clin. Genet. 15:418-429 (1979)
24. Mendelsohn ML, Hungerford DA, Mayall BH, Perry BH, Conway TJ, Prewitt JMS. Computer-oriented analysis of human chromosomes II. Integrated optical density as a single parameter for karyotype analysis. Ann. N.Y. Acad. Sci. 157:376-392 (1969)

25.  Minsky M. in: Winston P (ed) The Psychology of Computer Vision (McGraw Hill, New York, 1975)

26.  Oosterlinck A, van Daele J, Dom F, Reynaerts A, van den Berghe H. Computer aided karyotyping of human chromosomes. Proc. IEEE Conf. on Patt. Recog. Image Proc, Troy, NY, pp.61-69 (1977)

27.  Piper J, Granum E, Rutovitz D, Ruttledge H. Automation of chromosome analysis. Signal Processing 2:203-221 (1980)

28.  Piper J. Finding chromosome centromeres using boundary and density information. In: Simon J-C, Haralic RM (eds). Digital Image Processing (D. Reidel, Dordrecht, Netherlands, 1981) pp.511-518

29.  Piper J, Rutovitz D. Data structures for image processing in a C language and Unix environment. Patt. Rec. Letts. 3:119-129 (1985)

30.  Piper J. Classification of chromosomes constrained by expected class size. Patt. Rec. Letts. 4:391-395 (1986)

31.  Piper J, Towers S, Baldock R, Rutovitz D. Knowledge-Based Control of a Chromosome Analysis System. MRC Human Genetics Unit, internal report (1987)

32.  Piper J, Towers S, Gordon J, Ireland J, MacDougall D. Hypothesis combination and context sensitive classification for chromosome aberration scoring. In: Gelsema E, Kanal L (eds). Pattern Recognition in Practise III (Elsevier, Amsterdam, in press)

33.  Piper J, Granum E. On fully automatic feature measurement for banded chromosome classification. Cytometry, in press.

34.  Russell GT, Lane DM. A Knowledge-Based System for Enviromental Perception in a Subsea Context. IEEE J. of Oceanic Eng. OE 11:401-412(1986)

35.  Rutovitz D. Centromere finding: some shape descriptors for small chromosome outlines. in: Meltzer B, Michie D (eds) Machine Intelligence 5 (Edinburgh University Press, 1969) pp.435-462

36.  Rutovitz D. Chromosome classification and segmentation as exercises in knowing what to expect. In: Elcock EW, Michie D (eds) Machine Intelligence 8 (Ellis Horwood, Chichester, 1977) pp.455-472

37.  System Designers plc. POPLOG Users Guide (System Designers plc, 1986)

38.  Thomason MG, Granum E. Dynamically programmed inference of Markov networks from finite sets of sample strings. IEEE PAMI 8:491-501 (1986)

39.  Thornham A, Taylor CJ, Cooper DH. Object Cues for Model BasedImage Interpretation. Proc. 4th Alvey Vision Conf., Manchester, pp.53-58 (1988)

40.  Towers SJ. Frames as Data Structures for SBS, MRC Human Genetics Unit, internal report (1987)

41.  Vanderheydt L, Oosterlinck A, van den Berghe H. An application of fuzzy subset theory to the classification of human chromosomes. Proc. IEEE Conf. on Patt. Recog. and Image Processing, Chicago, Il, pp.466-472 (1979)

42.  Vanderheydt L, Dom F, Oosterlinck A, van den Berghe H. Two dimensional shape decomposition using fuzzy subset theory applied to automatic chromosome analysis. Pattern Recognition 13:147-157 (1981)

43.  van Vliet LJ, Young IT, ten Kate TK, Mayall BH, Groen FCA, Roos R. Athena: A Macintosh-Based Interactive Karyotyping System. (this volume)

44.  Vossepoel A. Analysis of image segmentation for automated chromosome identification. PhD Thesis, University of Leiden (1987)

43.  Wu Q, Suetens P, Oosterlinck A. Toward an expert system for chromosome analysis. Knowledge-based systems 1:43-52 (1987)

45.  Wu Q, Snellings J, Amory L, Suetens P, Oosterlinck A. A Polygonal Approximation Approach to Model-Based Contour Analysis in a Chromosome Segmentation System. (this volume)

46.  Zimmerman SO, Johnston DA, Arrighi FE, Rupp ME. Automated homologue matching of human G-banded chromosomes. Comput. Biol. Med. 16:223-233 (1986)

# Bibliography – Cytogenetics Automation 1975-88

## 1. Reviews and surveys

Bender MA: Unmet needs in automated cytogenetics. In: Automation of Cytogenetics. Mendelsohn ML. Ed. Asilomar Workshop. Lawrence Livermore Laboratory Technical report CONF-751158, 1975:pp.178-181.

Habbema JDF: Statistical Methods for Classification of Human Chromosomes. Biometrics 1979; 35:103-118.

Lubs HA: The current state of the art: Summary of discussion. In: Automation of Cytogenetics. Mendelsohn ML. Ed. Asilomar Workshop. Lawrence Livermore Laboratory Technical report CONF-751158, 1975:pp.174-177.

Lundsteen C, Philip J: Present status of automated chromosome analysis. In: Genetic Damage in Man Caused by Environmental Agents. Berg K. Ed. Academic Press, New York, 1979:pp.125-41.

Lundsteen C, Martin AO: On the selection of systems for automated cytogenetics. Am J Med Genet. In press.(1989).

Mendelsohn ML: Priorities for development and application of automated cytogenetics. In: Automation of cytogenetics. Mendelsohn ML. Ed. Asilomar Workshop. Lawrence Livermore Laboratory Technical report CONF-751158, 1975:pp.182-187.

Piper J, Granum E, Rutovitz D, Ruttledge H: Automation of chromosome analysis. Sign Proces 1980; 2:203-221.

Piper J: Chromosome Analysis: Image Processing or Pattern Recognition? In: Computing in Medicine. Paul JP, Jordan MM, Ferguson PMW, Andrews BJ. Eds. Macmillan, London 1982; pp.205-212.

Piper J, Lundsteen C: Human chromosome analysis by machine. Trends in Genet 1987; 3:309-313.

Rutovitz D: Automatic Chromosome Analysis. Pathologica 1983; 75 Suppl:210-242.

Stebbings JH: The relevance of automation to organizational problems in cytogenetic epidemiology. In: Automation of cytogenetics. Mendelsohn ML. Ed. Asilomar Workshop. Lawrence Livermore Laboratory Technical report CONF-751158, 1975:pp.170-173.

## 2. Automatic metaphase finding (algorithms, instrumentation, autofocus)

van den Berghe HTCM, France HF, Habbema JDF, Raatgever JW: Automated Selection of Metaphase Cells by Quality. Cytometry 1981; 1:363-368.

van Daele J, Dom F, de Buysscher L, Oosterlinck A, van den Berghe H: Metaphase finding using microprocessors. Proc IEEE Comp Soc Conf PRIP, Chicago, August 6-8. 1979:pp.460-465.

Farrow ASJ, Green DK, Rutovitz D: A cytogeneticist's microscope and a proposed system for aberration scoring. In: Automation of Cytogenetics. Mendelsohn ML. Ed. Asilomar Workshop. Lawrence Livermore Laboratory Technical report CONF-751158, 1975:pp.68-71.

Finnon P, Lloyd DC, Edwards AA: An assessment of the metaphase finding capability of the Cytoscan 110. Mutat Res 1986; 164:101-108.

Graham J, Pycock D: Automation of Routine Clinical Chromosome Analysis. II. Metaphase Finding. Anal Quant Cytol Histol 1987; 9:391-397.

Green DK, Bayley R, Rutovitz D: A cytogeneticist's Microscope. Microsc Acta 1977; 79:237-245.

Groen FCA, Young IT, Ligthart G: A comparison of different focus functions for use in autofocus algorithms. Cytometry 1985; 6:81-91.

Hutzler P: Preparation and evaluation of object spectra for automatic finding of chomosome metaphase spreads. Proc ICO-11 Conf. Madrid, Spain, Sept. 1979.

Lloyd D, Piper J, Rutovitz D, Shippey G: Multiprocessing interval processor for automated cytogenetics. Appl Optics 1987; 26:3356-3366.

Mason DC, Green DK: Automatic focusing of a computer controlled microscope. IEEE Trans Biomed Eng, BME-22 1975; 4:312-317.

Mason DC, Green DK: A computerised microscope focusing technique. Microsc Acta 1976; 78:439-448.

Rutovitz D, Green DK, Farrow ASJ, Mason DC: Computer assisted measurement in the cytogenetic laboratory. In: Pattern Recognition. Batchelor BG. Ed. Plenum 1978:pp.303-329.

Ruttledge H: The performance of an automatic metaphase finder used for chromosome aberration scoring. In: Machine-Aided Image Analysis. Gardner WE. Ed. Inst Phys London Conf Ser 44, 1979:pp.210-219.

Sabrin HW, Martin AO, Shaunnessey MS: Current status of metaphase locating devices: Preliminary clinical evaluation of one system. In: Proceedings of Computer Applications in Medical Care, Conference Nov. 1-4, Washington D.C, IEEE Computer Society Press Order, No. 377, 1981:pp.1100-1111.

Shafer DA, Mandelberg KI, Falek A: Computer Automation of Metaphase Finding, Sister Chromatid Exchange, and Chromosome Damage Analysis. Chem Mutag 1986; 10:357-380.

Shippey GA, Bayley RJH, Farrow ASJ, Rutovitz D, Tucker JH: A fast interval processor. Patt Recog 1981; 14:345-356.

Shippey G, Bayley R, Granum E: Metaphase finding using a fast interval processor, Metafip. Proceedings of the First IEEE Computer Society International Symposium on Medical Images and Image Interpretation, Berlin October 26-28, 1982:pp.349-357.

Shippey G: An automatic focussing device. US Patent 4636051 1984 (and UK Patent appliction 8409855).

Shippey G, Carothers AD, Gordon J: Operation and Performance of an Automatic Metaphase Finder Based on the MRC Fast Interval Processor. J Histochem Cytochem 1986; 34:1245-1252.

Vanderheydt L, Dom F, van Daele J, Jansen P, Oosterlinck A, van den Berghe H: Automatic search and quality assignment of metaphase of cell cultures. Proc. 5th. Nordic Meeting on Med. & Biol Engin., Linkoeping Sweden, June 1981; 2:pp.530-532.

# 3. Automatic chromosome classification (karyotyping, aberration scoring, pattern recognition, instrumentation, chromosome measurements, sister chromatid exchanges, dicentrics, centromere detection, image enhancement)

Aggarwal RK, Fu KS: A Pattern Classification System for the Identification of Irradiated Chromosomes. IEEE Trans Biom Eng BME-24 1977; 178-185.

Amory L: Conversion and improvement of a syntactical pattern recognition package for the resegmentation of metaphase images. Thesis, Catholic University of Leuven, 1988.

van den Berghe HTCM, Habbema JDF, France HF, Bakker HK, Vries GA: On the use of distribution-functions to describe and classify chromosome density-profiles. Comput Biol Med 1979; 9:11-20.

Bishop RP, Young IT: The automated classification of mitotic phase for human chromsome spreads. J Histochem Cytochem 1977; 25:730-740.

Carothers AD, Rutovitz D, Granum E: An Efficient Multiple-Cell Approach to automatic aneuploidy screening. Anal Quant Cytol 1983; 5:194-200.

Carayannopolous GL: An algorithm for segmentation of metaphase spreads. Patt Recog 1976; 8:151-161.

Drets ME: Bandscan, a computer program for on-line linear scanning of human banded chromosomes. Comput Programs Biomed 1978; 8:283-294.

Fantes J, Gosden J, Piper J: Use of an alphoid satellite sequence to locate the X chromosome automatically, with particular reference to the identification of fragile X. Cytogenet and Cell Genet 1988, in press.

Graham J, Taylor CJ, Cooper DH, Dixon RN: A compact set of image processing primitives and their role in a sucessful application program. Patt Recog Letts 1986; 4:325-333.

Granlund GH: Identification of Human Chromosomes Using Integrated Density Profiles. IEEE Trans Biomed Eng 1976; BME-23; 3:182-192.

Granlund GH, Zack GW, Young IT, Eden M: A technique for multiple-cell chromosome karyotyping. J Histochem Cytochem 1976; 24:160-167.

Granlund GH: The structure of a system for multiple-cell chromosome karyotyping. Proc. 4th Int. Joint Conf. on Pattern Recognition, Kyoto, Japan, Nov. 7-10, 1978:pp.837-841.

Granum E: Pattern Recognition aspects of Chromosome Analysis. Computerised and Visual Interpretation of Banded Human Chromosomes. Thesis, Electronics Laboratory, The Technical University of Denmark, Lyngby, December 1980.

Granum E: Application of statistical and syntactical methods of analysis and classification to chromosome data. In: Pattern Recognition Theory and Applications. Kittler J, Fu KS, Pau LF. Eds. NATO ASI (Oxford 1981), Reidel, Dordrecht, 1982; pp.373-398.

Granum E, Shippey G, Bayley R, Hamilton G, Rutovitz D: Real time digital thresholding of data from continuous Scanning Linear Arrays. Sign Proces 1987; 12:349-362.

Groen FC, Verbeek PW, Zee GA, Oosterlinck A: Some aspects concerning computation of chromosome banding profiles. In Proceedings of the 3rd International Joint Conference on Pattern Recognition, Coronado, California, 1976:pp.547-550.

Habbema JDF: A discriminant Analysis Approach to the Identification of Human Chromosomes. Biometrics 1976; 32:919-928.

Hazout S, Venuat AM, Valleron AJ, Rosenfeld C: Computer-Aided analysis of Chromosomal Aberrations Occuring in an Abnormal Human Karyotype. Hum Genet 1979; 49:133-145.

Hazout S, Mignit J, Guiguet M, Valleron AJ: Rectification of Distorted Chromosome Image: Automatic Determination of Density Profiles. Comput Biol Med 1984; 14:63-76

He L, Chai Z: Automatic detection of sister chromatid exchange. Proc. 9th Int. Conf. Pattern Recognition. 1988; pp.652-654

Ing PS, Ledley RS, Lubs HA: Chromosome analysis at the National Biomedical Research Foundation. In: Automation of Cytogenetics. Mendelsohn ML. Ed. Asilomar Workshop. Lawrence Livermore Laboratory Technical report CONF-751158, 1975:pp.27-38.

Janardan KG, Schaeffer DJ: Models for the Analysis of Chromosomal Aberrations in human Leukocytes. Biom J 1977; 19:599-612.

Ji L: Intelligent Splitting in the Chromosome Domain. Patt Recog 1989; in press.

Ledley RS, Ing PS, Lubs HA: Human chromosome classification using discriminant analysis and Bayesian probability. Comput Biol Med 1980; 10:209-218.

Lee ET: Algorithms for finding most dissimilar images with possible applications to chromosome classification. Bull Math Biol 1976; 38:505-517.

Lockwood DH, Riccardi VM, Zimmermann SO, Johnston DA: Prophase chromosomes unique band sequences: Definition and utilization. Cytogenet Cell Genet 1986; 42:141-153.

Loos P, Cremer T, Hausmann M, Jauch A, Emmerich P, Schlegel W, Cremer C: Distances between two chromosomes in interphase nuclei as determined with digital image analysis. In: Clinical Cytometry and Histometry. Burger G, Ploem JS, Goerttler K. Eds. Academic Press 1987:pp.285-287.

Lundsteen C, Lind AM, Granum E: Visual classification of banded human chromosomes I. Karyotyping compared with classification of isolated chromosomes. Ann Hum Genet (Lond.) 1976; 40:87-97.

Lundsteen C, Granum E: Visual classification of banded human chromosomes II. Classification and karyotyping of integrated density profiles. Ann Hum Genet (Lond.) 1977; 40:431-42.

Lundsteen C, Granum E: Description of chromosome banding patterns by band transition sequences. Clin Genet 1979, 15:418-29.

Lundsteen C, Granum E: Visual classification of banded human chromosomes III. Classification and karyotyping of density profiles described by band transition sequences. Clin Genet 1979; 15:430-39.

Lundsteen C: Aspects of automated chromosome analysis. Different representations of banded human chromosomes and their cytogenetic evaluation. Thesis. Dan Med Bull 1980; 17:183-90.

Lundsteen C, Philip J, Granum E: Quantitative analysis of 6985 digitized trypsin G-banded human metaphase chromosomes. Clin Genet 1980; 18:355-70.

Lundsteen C, Bjerregaard B, Granum E, Philip J, Philip K: Automatic chromosome analysis I. A simple method for classification of B- and D-group chromosomes represented by band transition sequences. Clin Genet 1980; 17:183-90.

Lundsteen C, Gerdes T, Granum E, Philip J, Philip K: Automatic chromosome analysis II. 6989 chromosomes described by band transition sequences karyotyped into 24 classes with an error rate of 3%. Clin Genet 1981; 19:26-36.

Mason DC, Lauder I, Rutovitz D, Spowart G: Measurement of C-bands in human chromosomes. Comput Biol Med 1975; 5:179-201.

Mason DC, Rutovitz D: The economics of automated aberration scoring. In: Mutagen induced chromosome aberrations. Evans HJ, Lloyd DC. Eds. Edinburgh University Press, 1978:pp.339-345.

Mayall BH, Carrano AV, Golbus MS, Conte FA, Epstein CJ: DNA cytophotometry in prenatal cytogenetic diagnosis. Clin Genet 1977; 11:273-276.

Mayall BH, Carrano AV, Moore DH, Rowley JD: Quantification by DNA-based cytophotometry of the 9q+/22q-chromosomal translocation associated with chronic myelogeneous leukemia. Cancer Res 1977; 37:3590-3593.

Mayall BH, Carrano AV, Moore DH, Ashworth LK, Bennet DE, Mendelsohn ML: The DNA-based human karyotype. Cytometry 1984; 5:376-385.

Moore DH II: Normalisation of chromosome measurements: A new method. Comput Biol Med 1975; 5:21-28.

Moore RC, Woodward A, Randell C, Koschel K, Hodgson GS: Quantitative identification of banded human chromosomes. Clin Genet 1982; 21:411-417.

Neurath PW, Gallus G, Horton JB, Selles W: Automatic karyotyping: Progress, Perspectives and Economics. In: Automation of Cytogenetics. Mendelsohn ML. Ed. Asilomar Workshop. Lawrence Livermore Laboratory Technical report CONF-751158, 1975:pp.17-26.

Nickolls P, Piper J, Rutovitz D, Chisholm A, Johnstone I, Robertson M: Pre-processing of images in an automated chromosome analysis system. Patt Recog 1981; 14:219-229.

Oosterlinck A, van Daele J, Boer J, Dom F, Reynaerts A, van den Berghe H: Computer-assisted karyotyping with human interaction. J Histochem Cytochem 1977; 25:754-762.

Oosterlinck A, van Daele J, Dom F, Reynaerts A, van den Berghe H: Computer aided karyotyping of human chromosomes. Proc. IEEE Conf on Pattern Recognition and Image Processing Troy, NY, 1977:pp.61-69.

van der Ploeg M, Vossepoel AM, Bosman FT, van Duijn P: High-resolution scannong-densitometry of photographic negatives of human chromosomes; III. Determination of fluorescence emission intensities. Histochemistry 1977; 51:269-291.

Piper J: Finding chromosome centromeres using boundary and density information. In: Digital Image Processing. Simon J-C, Haralick RM. Eds. D. Reidel, Dordrecht, Netherlands 1981:pp.511-518.

Piper J: Interactive image enhancement and analysis of prometaphase chromosomes and their band patterns. Anal Quant Cytol 1982; 4:223-240.

Piper J, Nickolls P, McLaren W, Rutovitz D, Chisholm A, Johnstone I: The effect of digital image filtering on the performance of an automatic chromosome classifier. Sign Proces 1982; 4:361-373.

Piper J: Image Restoration and Interactive Processing for Analysis of High Resolution Banded Chromosomes. Proc 1st Int. Symposium on Medical Imaging and Image Interpretation, Berlin 1982:pp.24-29.

Piper J, Elder JK: Image enhancement in biological microscopy. Proc. Int. Conf. Image Assessment: Infrared and Visible. Oxford, December 1983:pp.12-14.

Piper J: Classification of chromosomes constrained by expected class size. Patt Recog Letts 1986; 4:391-395.

Piper J: The effect of zero feature correlation assumption on maximum likelihood based classification of chromosomes. Sign Proces 1987; 12:49-57.

Piper J, Granum E: On fully automatic feature measurement for banded chromosome classification. Cytometry 1989, in press.

Piper J, Towers S, Gordon J, Ireland J, McDougall D: Hypothesis combination and context sensitive classification for chromosome aberration scoring. In: Pattern Recognition and Artificial Intelligence. Gelsema ES, Kanal LN. Eds. Elsevier, Amsterdam 1988:pp.449-460.

Rutovitz D: Chromosome classification and segmentation as exercises in knowing what to expect. In: Machine Intelligence 8. Elcock EW, Michie D. Eds. Ellis Horwood, Chichester, UK 1977:pp.455-472.

Sekiya T, Saito M, Ikeda K: Chromosome classification system based on Banding Technique. Med Progr Technol 1979; 6:169-177.

Selles WD, Neurath PW, Conklin J, Horton JB: Progress and problems in the cy-tophotometric analysis of G-banded chromosomes. In: Automation of Cytogenet-ics. Mendelsohn ML. Ed. Asilomar Workshop. Lawrence Livermore Laboratory Technical report CONF-751158, 1975:pp.3-17.

Shepherd B, Piper J, Rutovitz D: Comparison of ACLS and classical linear methods in a biological application. In: Machine Intelligence 11. Hayes JE, Michie D, Richards J. Eds. Oxford University Press 1988:pp.423-434.

Slot RE: On the profit of taking into account the known number of objects per class in classification methods. IEEE Trans Inform Theory 1979; 25:484-488.

Snellings J: Segmentation and feature extraction - A syntactical pattern recognition approach to the preprocessing of metaphase images. Thesis, Catholic University of Leuven, 1987.

Thomason MG, Granum E: Dynamic Programming Inference of Markov Networks from Finite Sets of Sample Strings. IEEE-transactions on Pattern Analysis and Machine Intelligence (PAMI) 1986; 8:491-501.

Thomason MG, Granum E: Sequential Inference of Markov Networks by Dynamic Programming for Structural Pattern Recognition. Patt Recog Letts 1987; 5:31-39.

Tso MKS, Graham J: The transportation algorithm as an aid to chromosome classification. Patt Recog Letts 1983; 1:489-496.

Vanderheydt L, Oosterlinck A, van den Berghe H: Design of a special Interpreter for the classification of human chromosomes. IEEE transactions on Pattern analysis and Machine Intelligence Special Issue, Vol. PAMI-I, No. 2, April 1979:214-219.

Vanderheydt L, Dom F, Oosterlinck A, van den Berghe H: An application of fuzzy subset theory to the classification of human chromosomes. Proc IEEE Conf on Pattern Recognition and Image Processing Chicago II 1979:pp.466-472.

Vanderheydt L, Oosterlinck A, van Daele J, van den Berghe H: Design of a graph-representation and a fuzzy-classifier for human chromosomes. Patt Recog 1980; 12: 201-210.

Vanderheydt L, Dom F, Oosterlinck A, van den Berghe H: Decomposition of touching chromosomes using fuzzy subset theory. Proc. 1st. Scan Conf on Image Analysis, Linkoeping, Sweden January 14-16, 1980:pp.348-354.

Vanderheydt L, Dom F, Oosterlinck A, van den Berghe H: Two-dimensional shape decomposition using fuzzy subset theory applied on automated chromosome analysis. Patt Recog 1981; 13:147-157.

Vossepoel AM: DIODA: delineation and feature extraction of microscopical objects. Comput Programs Biomed 1979; 19:231-244.

Vossepoel AM: Application of some local segmentation methods in chromosome analysis. In: First quinquennial review 1981-86, Dutch Society for Pattern Recognition. Back E, Duin RPW, Gelsema ES, Kamminga C, Valeton JM, Vossepoel AM. Eds. Pijnacker: D.E.B. 1986; pp.135-142.

Vossepoel AM: Analysis of Image Segmentation for Automated Chromosome Identification. Thesis, University of Leiden, 1987.

Wayne AW, Sharp JC: The use of high resolution microscope photometry in the discrimination of chromosome bands. J Microsc 1981; 124:163-167.

Willborn K, Cremer C, Hausmann M, Cremer C: An approach to analysis of structural aberrations on the basis of schematic representations of chromosomes by a personal computer. Eur J Cell Biol 1987; 43(Suppl 17):64.

Wu Q, Suetens P, Oosterlinck A: Toward an expert system for chromosome analysis. Knowledge-Based Systems 1987; 1:43-52.

Wu Q, Suetens P, Oosterlinck A, van den Berghe H. An expert image analysis system for chromosome analysis application. In: Medical Imaging, Session 12: Imaging Processing III, 1-6 February, 1987. Proc. S.P.I.E.; 767:pp.400-405.

Zack GW, Spriet JA, Latt SA, Granlund GH, Young IT: Automatic detection and localization of sister chromatid exchanges. J Histochem Cytochem 1976; 24:168-177.

Zack GW, Rogers WE, Latt SA: Automatic measurement of Sister Chromatid Exchange frequency. J Histochem Cytochem 1977; 25:741-753.

Zimmermann SO, Johnston DA: Automated cytogenetics at the M.D. Anderson Hospital. In: Automation of Cytogenetics. Mendelsohn ML. Ed. Asilomar Workshop. Lawrence Livermore Laboratory Technical report CONF-751158, 1975:pp.85-95.

Zimmermann SO, Johnston DA, Arrighi FE, Rupp ME: Automated homologue matching of human G-banded Chromosomes. Comput Biol Med 1986; 16:223-233.

# 4. Systems for automated cytogenetic analysis (instrumentation, clinical cytogenetics, aberration scoring, research systems)

Bille J, Scharfenberg H, Maenner R: Biological dosimetry by chromosome aberration scoring with parallel image processing with the Heidelberg Polyp polyprocessor system. Comput Biol Med 1983; 13:49-79.

Bille J, Loerch T, Stephan G, Wittler C: Automated cytogenetics dosimetry. Proc IEEE 9th Annual Conf. Eng. in Med and Biol. Boston Nov. 13-16, 1987, pp.1157-1158

Bruschi C, Tedeschi F, Puglisi PP, Marmiroli N: Computer-assisted karyotyping system of banded chromosomes. Cytogenet Cell Genet 1981; 29:1-8.

Caspersson T: TV-based tools for rapid analysis of chromosome banding patterns. In: Automation of Cytogenetics. Mendelsohn ML. Ed. Asilomar Workshop. Lawrence Livermore Laboratory Technical report CONF-751158, 1975:pp.122-131.

Castleman KR, Melnyk J, Frieden HJ, Persinger GR, Wall RJ: A minicomputer based karyotyping system. In: Automation of Cytogenetics. Mendelsohn ML. Ed. Asilomar Workshop. Lawrence Livermore Laboratory Technical report CONF-751158, 1975:pp.46-50.

Castleman KR, Melnyk JH: Automatic system for chromosome analysis-final report, JPL Document No 5040-30. Jet Propulsion Laboratory, Pasadena, California, 1976.

Castleman KR, Melnyk J, Frieden HJ, Persinger GW, Wall RJ: Computer-assisted karyotyping. J Reprod Med 1976; 17:53-57.

Castleman KR, Melnyk J, Frieden HJ, Persinger GW, Wall RJ: Karyotype analysis by computer and its application to mutagenicity testing of environmental chemicals. Mutat Res 1976; 41:153-162.

Dewald GW, Robb RA, Gordon H: A computer-based videodensitometric system for studying banded human chromosomes illustrated by the analysis of the normal morphology of chromosome 18. Am J Hum Genet 1977; 29:37-51.

Farrow ASJ, Green DK, Rutovitz D: A cytogeneticist's microscope and a proposed system for aberration scoring. In: Automation of Cytogenetics. Mendelsohn ML. Ed. Asilomar Workshop. Lawrence Livermore Laboratory Technical report CONF-751158, 1975:pp.68-71.

Fontana G, Matteuzzi P, Fabiani G, Forabosco A: Interactive system for automatic analysis of human R-band metaphases. Acta Genet Med Gemellol 1975; 24:299-306.

Geneix A, Malet P, Bonton P, Grouche L, Perissel B: Image Processing in Human Cytogenetics: New steps toward quantification. Karyogram (USA), 1988; 14:45-49.

Gilbert CW, Muldal S: Measurement and computer system for karyotyping human and other cells. Nature New Biol 1977; 230:203-207.

Graham J, Taylor CJ. Automated chromosome analysis using the Magiscan image analyser. Anal Quant Cytol 1980; 2:237-242.

Graham J: Automation of routine clinical chromosome analysis. I. Karyotyping by machine. Anal Quant Cytol Histol 1987; 9:383-390.

Graham J, Pycock D: Automation of Routine Clinical chromosome analysis. II. Metaphase Finding. Anal Quant Cytol Histol 1987; 9:391-397.

Green DK, Bayley R, Rutovitz D: A cytogeneticist's microscope. Microsc Acta 1977; 79:237-245.

Kasvand T, Hamill P, Bora KC, Douglas G: Experimental online karyotyping at the National Research Council of Canada. In: Automation of Cytogenetics. Mendelsohn ML. Ed. Asilomar Workshop. Lawrence Livermore Laboratory Technical report CONF-751158, 1975:pp.96-109.

Kliment V, Tuscany R, Dvorak J, Tomasek L: Computer-assisted analysis of human G-banded chromosomes. Comput Biomed Res 1983; 16:20-28.

Ledley RS, Buas M, Golab TL, Arminski L. Chromoscan: the new instrumentation for cytogenetics. Proc IEEE 9th Annual Conf. Eng. in Med and Biol. Boston Nov. 13-16, 1987, pp.1159-

Le Go R: Automation in cytogenetics at CEA Paris. In: Biomedical images and computers. Sklansky J, Biscarde JC. Eds. Springer-Verlag, 1982; pp.66-77.

Leonard C, Saint-Jean P, Schoevaert D, Eydoux P, Girard S, LeGo R: An automatic system for chromosomal analysis applied to prenatal diagnosis. Hum Genet 1979; 47:319-327.

Lloyd D, Piper J, Rutovitz D, Shippey G: A multiprocessing interval processor for automated cytogenetics. Appl Optics 1987; 26:3356-3366.

Loerch T, Bille J, Frieben M, Stephan G: An automated biological dosimetry system. SPIE Proc 1985; 596:199-206.

Lundsteen C, Gerdes T, Philip J, Graham J, Pycock D: An interactive system for chromosome analysis. Tests of clinical performance. Proc 3rd Scan Conf Image Analysis, Copenhagen, 1983; pp.392-397.

Lundsteen C, Gerdes T, Maahr J: Automatic classification of chromosomes as part of a routine system for clinical analysis. Cytometry 1986; 7:1-7.

Lundsteen C, Gerdes T, Maahr J, Philip J: Clinical performance of a routine system for semi-automated chromosome analysis. Am J Hum Genet 1987; 41:493-502.

Mayall BH, Carrano AV, Moore DH II, Ashworth LK, Bennett DE, Bogart E, Littlepage JL, Minkler JL, Piluso DL, Mendelsohn ML: Cytophometric analysis of human chromosomes. In: Automation of Cytogenetics. Mendelsohn ML. Ed. Asilomar Workshop. Lawrence Livermore Laboratory Technical report CONF-751158, 1975:pp.135-144.

Oosterlinck A, Vlietinck RF: Characteristics of an automated clinical cytogenetics system. In: Automation of Cytogenetics. Mendelsohn ML. Ed. Asilomar Workshop. Lawrence Livermore Laboratory Technical report CONF-751158, 1975:pp.72-84.

Oosterlinck A, van Daele J, Boer F, Reynaerts DA, van den Berghe H: Computer-assisted karyotyping with human interaction. J Histochem Cytochem 1977; 25:754-762.

Picciano D, Kilian DJ: A semi-automated cytogenetic analysis system. Exp Cell Res 1977; 107:431-435.

Piper J, Breckon G: An automated system for karyoyping mouse chromosomes. Cytog Cell Genet 1989; in press

Piper J, Mason D, Rutovitz D, Ruttledge H, Smith L: Efficient interaction for automated chromosome analysis using asynchronous parallel processes. J Histochem Cytochem 1979; 27:432-435.

Piper J, McDermott R: Asynchronous parallel processing for interactive automated chromosome analysis. In: Machine-Aided Image Analysis, Inst Phys Conf Ser 44. Gardner WE. Ed. The Institute of Physics, London 1979:pp.195-202.

Piper J, Rutovitz D: A parallel processor implementation of a chromosome analysis system. Patt Recog 1986; 4:397-404.

Rutovitz D, Piper J: The balance of special and conventional computer architecture in an image processing application. In: Multicomputers and Image Processing Algorithms and Programs. Preston K, Uhr L. Eds. Academic Press, New York, 1982:pp.161-177.

Sekiya T, Saito M. Automatic system for identifying banded human chromosomes. Proc Int Symposium Medical Information Systems (Medis 78) Osaka 1978, pp.176-178.

Shaunnessey M, Martin AO, Rzeszotarski M, Thomas CW, Isenstein B: Status of a computer-aided system for cytogenetic specimen preparation and analysis. In: Proceedings of the Third International IEEE COMPSAC, 1979:pp.276-281.

Taylor CJ, Graham J, Cooper D: System architectures for interactive knowledge-based image interpretation. Phil Trans Roy Soc Lond 1988; A324:451-465.

Taylor CJ, Dixon RN, Gregory PJ, Graham J: An architecture for integrating symbolic and numeric image processing. In: Intermediate-level image processing. Duff MJB. Ed. Academic Press, London, 1986; pp.19-34.

Ten Kate TK: Design and implementation of an interactive karyotyping program in C on a Vicom digital image processor. Thesis, Department of Applied Physics, Delft University of Technology 1985.

Vanderheydt L, de Roo J, Oosterlinck A, van den Berghe H: Licas: Leuven Interactive chromosome analysis system. Proc. 2nd Scand. Conf. on Image Analysis, June 15-17, 1981, Helsinki, Finland:pp.393-398.

Vrolijk J, Brinke H, Ploem JS, Pearson PL: Video techniques applied to chromosome analysis. Microsc Acta 1980; Suppl 4:108-115.

Vrolijk J, Pearson PL, Ploem JS: Leytas, a system for the processing of microscopic images. Anal and Quant Cytol 1980; 2:41-48.

Vrolijk J, Pearson PL, Ploem JS: TAL: An interpretive language for the Leyden Television Analysis System. In: Languages and Architectures for Image Processing. Duff MJB, Levialdi S. Eds. Academic Press 1981:pp.125-137.

Wakazono S, Ichihara N, Hibi K, Ariga T: The development of chromosome image data processing and executer system. Proc Int Symposium Medical Information Systems (Medis 78) Osaka 1978, pp.174-176.

Wald N, Li CC, Herron JM, Davis L, Fatora SR: Automated Analysis of Chromosome damage. In:Automation of Cytogenetics. Mendelsohn ML. Ed. Asilomar Workshop. Lawrence Livermore Laboratory Technical report CONF-751158, 1975:pp.39-45.

Wald N, Fatora SR, Herron JM, Preston K, Li CC, Davis L: Status report on automated chromosome aberration detection. J Histochem Cytochem 1976; 24:156-159.

Wienberg J, Maurer A, Berneis M: Image Processing in Cytogenetic Routine Diagnostics and Basic Research. Internat J Anthropol 1988; 3:51-61.

Wulf HC: Operator-assisted semi-automatic karyotyping of banded metaphases. Cytogenet Cell Genet 1977; 19:146-158.

Wulf HC: Semi-Automatic karyotyping with Operator-Classification of Banded chromosomes. Mikroskopie 1984; 41:227-235.

Wulf HC, Philip J: Semi-automatic karyotyping facility - a clinical test. Hereditas 1986; 105:37-40.

Zack GW: An automatic chromosome image processing system for multiple-cell karyotyping and sister chromatid exchange analysis. In: Automation of Cytogenetics. Mendelsohn ML. Ed. Asilomar Workshop. Lawrence Livermore Laboratory Technical report CONF-751158, 1975:pp.131-134.

# 5. Chromosome analysis by flow cytometry

Arkesteijn GJ, van Dierendonck JH, Vossepoel Am, Cornelisse CJ: Flow karyotyping of human melanome cell lines. Cytometry 1986; 7:425-430.

Aten JA, Kooi MW, Bijman JTH, Kipp JBA, Barensden GW: Flow cytometric analysis of chromosome damage after irradiation: Relation to chromosome aberrations and cell survival. In: Biological Dosimetry. Eisert WG, Mendelsohn ML. Eds. Springer-Verlag, 1984:pp.57-60.

Carrano AV, Gray JW, Langlois RG, Burkhart-Schultz KJ, van Dilla MA: Measurement and purification of human chromosomes by flow cytometry and sorting. Proc Natl Acad Sci USA 1979; 76:1382-1384.

Collard JG, Phillippus E, Tulp A, Lebo RV, Gray JW: Separation and analysis of human chromosomes by combined velocity sedimentation and flow cytometry. Cytometry, 1984:5:9-19.

Cooke A, Tolmie J, Darlington W, Boyd E, Thomson R, Ferguson-Smith MA: Confirmation of a suspected 16q deletion in a dysmorphic child by flow karyotype analysis. J Med Genet 1987; 24:88-92.

Cooke A, Gillard EF, Yates JRW, Mitchell MJ, Aitken DA, Weir DM. Affara NA, Ferguson-Smith MA: X chromosome deletions detectable by flow cytometry in some patients with steroid sulphatase deficiency (X-linked icthyosis). Human Genet 1988; 79:49-52.

Cremer C, Gray JW, Ropers HH: Flow cytometric characterization of a Chinese hamster x man hybrid cell line retaining the human Y chromosome. Hum Genet 1982:60:262-266.

Cremer C, Gray JW: Application of the BrdU/thymidine method to flow cytogenetics: Differential quenching/enhancement of the Hoechst 33258 fluorescence of late replicating chromosomes. Cell Genet 1982; 8:319-327.

Cremer C, Cremer T, Gray JW: Induction of chromosome damage by ultraviolet light and caffeine: Correlation of cytogenetic evaluation and flow karyotype. Cytometry 1982; 2:287-290.

Cremer C, Gray JW: Replication kinetics of Chinese hamster chromosomes as revealed by bivariate flow karyotyping. Cytometry 1983; 3:282-286.

Deaven LL: Karyotype analysis of chinese hamster chromosomes by flow microfluorometry. In: Automation of Cytogenetics. Mendelsohn ML. Ed. Asilomar Workshop. Lawrence Livermore Laboratory Technical report CONF-751158, 1975: pp.165-169.

Deaven LL, van-Dilla MA, Bartholdi MF, Carrano AV, Cram LS, Fuscoe JC, Gray JW, Hildebrand CE, Moyzis RK, Perlman J: Construction of human chromosome-specific DNA libraries from flow-sorted chromosomes. Cold Spring Harbor Symp Quant Biol, 1986; 51:159-167.

Dudin G, Hausmann M, Rens W, Aten J, Cremer C: Fluorescence hybridization of isolated metaphase chromosomes in suspension for flow cytometric analysis. Annales Universitatis Saravensis Medicinae 1987; Suppl.7:81-84.

Fantes JA, Green DK, Elder J, Malloy P, Evans HJ: Detecting radiation damage to human chromosomes by flow cytometry. Mutat Res 1983; 119:161-168.

Fantes JA, Green DK, Cooke HJ: Purifying Human Y chromosomes by flow cytometry and sorting. Cytometry 1983; 4:88-91.

Fantes JA, Green DK: Human Chromosome Analyses and sorting. In Methods in Molecular Biology. Pollard JW, Walker JM. Eds. Humana Press 1988.

Gray JW, Carrano AV, Steinmetz LL, van Dilla MA, Moore DH, Mayall BH, Mendelsohn ML: Chromosome measurement and sorting by flow systems. Proc Nat Acad Sci USA 1975; 72:1231-1234.

Gray JW, Carrano AV, Moore DH, Steinmetz LL, Minkler J, Mayall B, Mendelsohn ML, van Dilla MA: High speed quantitative karyotyping by flow microfluorometry. Clin Chem, 1975; 21:1258-1262.

Gray JW, Langlois RG, Carrano AV, Burkhart-Schulte K, van Dilla MA: High resolution chromosome analysis: One and Two Parameter Flow cytometry. Chromosoma 1979; 73:9-27.

Gray JW, Peters D, Merrill J, Martin R, van Dilla MA: Slit-scan flow cytometry of mammalian chromosomes. J Histochem Cytochem 1979; 27:441-444.

Gray JW, Langlois RG: Chromosome classification and purification using flow cytometry and sorting. Annu Rev Biophys Biophys Chem 1986; 15:195-235.

Gray JW, Dean PN, Fuscoe JC, Peters DC, Trask BJ, van den Engh GJ, van Dilla MA: High-speed chromosome sorting. Science 1987; 238:323-329.

Gray JW, Trask B, Engh G, Silva A, Lozes C, Grell S, Schonberg S, Yu LC, Golbus MS: Application of flow karyotyping in prenatal detection of chromosome aberrations. Am J Hum Genet 1988; 42:49-59.

Green DK, Fantes JA: Improved accuracy of in-flow chromosome fluorescence measurements by digital processing of multi-parameter flow data. Sign Proces 1982; 5: 175-186.

Green DK, Fantes JA, Buckton KE, Elder JK, Malloy P, Carothers A, Evans HJ: Karyotyping and identification of human chromosome polymorphisms by single fluorochrome flow cytometry. Hum Genet 1984; 66:143-146.

Green DK, Fantes JA, Spowart G: Radiation dosimetry using the methods of flow cytogenetics. In Biological Dosimetry, Eisert WG, Mendelsohn ML. Eds. Springer-Verlag 1984:pp.67-76.

Green DK, Fantes JA, Evans HJ: Human metaphase chromosomes: Analysis and sorting by flow cytometry. In Genetic Disorders of the Fetus (second edition). Milunsky A. Ed. Plenum, New York 1986.

Green DK, Fantes JA: Counting dicentric chromosomes by flow cytometry. In Clinical Cytometry and Histometry. Burger G, Ploem JS Goerttler K. Eds. Academic Press 1987:pp.263-268.

Harris P, Boyd E, Fergusson-Smith MA: Optimising human chromosome separation for the production of chromosome-specific libraries by flow sorting. Hum Genet 1985; 70:59-65.

Harris P, Cooke A, Boyd E, Young BD, Fergusson-Smith MA: The potential of family flow karyotyping for the detection of chromosome abnormalities. Hum Genet 1987; 76:129-133.

Harris P, Boyd E, Young BD, Ferguson-Smith MA: Determination of the DNA content of human chromosomes by flow cytometry. Cytogenet Cell Genet 1986; 41:14-21.

Hausmann M, Dudin G, Aten JA, Bier F, Cremer C: Slit - scan flow cytometry following fluorescence hybridization: A new approach to detect chromosome translocations. Eur J Cell Biol 1988; 46 (suppl. 22):26.

Krumlauf R, Jean Pierre R, Young BD: Construction and characterisation of genomic libraries for specific human chromosomes. Proc Natl Acad Sci USA 1982; 79:2971-2975.

Lebo RV, Gorin F, Flitteric RJ, Kao FT, Cheung MC, Bruce BD, Kan YW: High resolution chromosome sorting and DNA spot blot analysis assign McArdl's syndrome to chromosome 11. Science 1984; 225:57-59.

Lebo RV, Golbus MS, Cheung MC: Detecting abnormal human chromosome constitutions by dual laser flow cytogenetics. Am J Med Genet 1986; 25:519-529.

Lucas JN, Gray JW, Peters DC, van Dilla MA: Centromeric index measurement by slit-scan flow cytometry. Cytometry 1983; 4:109-116.

Lucas JN, Gray JW: Centromeric index versus DNA content flow karyotypes of human chromosomes measured by means of slit-scan flow cytometry. Cytometry 1987; 8:273-279.

Lucas JN, Lozes C, Mullikin D, Pinkel D, Gray J: Automated quantification of the frequency of aberrant chromosomes in human cells. Cytometry 1987; Suppl 1:13.

van Dilla MA, Carrano AV, Gray JW: Flow karyotyping: Current Status and Potential Development. In: Automation of cytogentics. Mendelsohn ML. Ed. Asilomar Workshop. Lawrence Livermore Laboratory Technical report CONF-751158, 1975:pp.145-164.

Mayall BH, Carrano AV, Moore DH, Ashworth L, Bennett DE, Mendelsohn ML: The DNA based human karyotype. Cytometry 1984; 5:376-385.

Wilcox DE, Cooke A, Colgan J, Boyd E, Aitken DA, Sinclair L, Glasgow L, Stephenson JBP, Ferguson-Smith MA: Duchenne muscular dystrophy due to familial Xp2.1 deletion detectable by DNA analysis and flow cytometry. Hum Genet 1986; 73:175-180.

Young BD, Ferguson-Smith MA, Sillar R, Boyd E: High-resolution analysis of human peripheral lymphocyte chromosomes by flow cytometry. Proc Natl Acad Sci USA 1981; 78:7727-7731.

# 6. Systems for automated preparation of chromosomes

Cimino MC, Martin AO, Maremont SM, Shaunnessey MS, Simpson JL: Automated cytogentic processing for human population monitoring. Progr Mutat Res 1982; 3:169-173.

Melnyk J, Garland PW, Mount B: A semi-automated specimen preparation system for cytogenetics. In: Automation of Cytogenetics. Mendelsohn ML. Ed. Asilomar Workshop. Lawrence Livermore Laboratory Technical report CONF-751158, 1975:pp.51-68.

Melnyk J, Garland PW, Porter W: Semiautomation of Chromosome Preparations. Birth Defects 1975; 11:277-282.

Melnyk J, Garland PW, Mount B: A Semi-Automated Specimen Preparation System for Cytogenetics. J Reprod Med 1976; 17:59-67.

Shaunnessey MS, Martin AO, Rzeszotarski M, Thomas CW, Isenstein B: Status of a computer-aided system for cytogenetic specimen preparation and analysis. Proceedings of third International Computer Software & Applications Conference, 1979:pp.276-281.

Shaunnessey MS, Martin AO, Sabrin HW, Cimino MC, Rissman A: A new era for cytogenetics laboratories: Automated specimen preparation. Proceedings of the fifth annual symposium on computer applications in medical care 1981:pp.538-541.

Spureck J, Carlson R, Allen J, Dewald G: Culturing and robotic harvesting of bone marrow, lymph nodes, peripheral bloods, solid tumors with in situ techniques. Cancer Genet Cytogenet 1988; 32:59-66.

Vrolijk J, Korthof G, Vletter G, van der Geest CRG, Gerese GW, Pearson PL: The automation of culturing and harvesting of human chromosome specimens. Histochemistry 1986; 84:586-593.

Vrolijk J, Korthof G, Vlettr G, van der Geest CRG, Gerese GW, Pearson PL: An automated culturing and harvesting system for cytogenetics. In: Clinical Cytometry and Histometry. Proceedings of the International Symposium on Clinical Cytometry and Histometry. Burger G, Ploem JS, Goerttler K. Eds. Academic press 1987; pp.139 -142.

Wulf HC: Mechanical Preparation of Cells for Chromosome Studies. In: Chromosomes today. Pearson PL, Lewis KR. Eds. Proceedings of the Leiden Chromosome Conference. John Wiley & Sons, New York 1974:pp.439-443.

Wulf HC: Mechanical Preparation of Cells for Chromosome Studies. Hum Hered 1975; 25:398-401.

# Subject Index